FORM AND MEANING IN LANGUAGE

CSLI Lecture Notes
Number 121

CHARLES J. FILLMORE

Form and Meaning in Language

Volume 1: Papers on Semantic Roles

CSLI Publications
Center for the Study of
Language and Information
Stanford, California

Center for the Study of Language and Information
Leland Stanford Junior University
07 06 05 04 03 5 4 3 2 1

Printed in the United States

Library of Congress Cataloging-in-Publication Data

Fillmore, Charles J.
Form and meaning in language : papers on semantic roles /
by Charles J. Fillmore.
p. cm.–(CSLI lecture notes ; v. 121)
Includes bibliographical references and index.

ISBN: 1-57586-285-9 (v. 1 : hardback : alk. paper)
ISBN: 1-57586-286-7 (v. 1 : pbk. : alk. paper)
1. Linguistics. I. Title. II. Series.
P121.F4786 2003
410–dc21 2002156118
CIP

∞ The acid-free paper used in this book meets the minimum requirements of the American National Standard for Information Sciences–Permanence of Paper for Printed Library Materials, ANSI Z39.48-1984.

CSLI was founded in 1983 by researchers from Stanford University, SRI International, and Xerox PARC to further the research and development of integrated theories of language, information, and computation. CSLI headquarters and CSLI Publications are located on the campus of Stanford University.

CSLI Publications reports new developments in the study of language, information, and computation. In addition to lecture notes, our publications include monographs, working papers, revised dissertations, and conference proceedings. Our aim is to make new results, ideas, and approaches available as quickly as possible. Please visit our web site at
http://cslipublications.stanford.edu
for comments on this and other titles, as well as for changes and corrections by the author and publisher.

Contents

Preface

The supportive community in which my writings on "case grammar" emerged is symbolized in my mind by a lunch table on the top floor of the old University Hall at Ohio State University in the mid-1960s, in the context of Bill Wang's "Project on Linguistic Analysis" and a very young linguistics program. Among those in attendance at many of these conversations were Bill Wang, Ilse Lehiste, Terry Langendoen, Greg Lee, Jim Heringer, Sandy Annear (Thompson), and the mathematician Leroy F. Meyers. The first audiences were my students in the new linguistics program there.

Although I had no direct training in the transformational generative grammar movement, I considered myself a faithful member of the club at the time, and the first few papers in this collection were intended as contributions to that tradition. In the end those parts of the larger linguistics world that paid attention to case grammar saw it as an alternative to transformational grammar.

Case grammar, so-called, emerged almost simultaneously with Generative Semantics, and the two activities moved in parallel, through friendship, but never merged. They had in common a plan to present almost everything that had to do with meaning in a single initial level of representation and to take care of everything else, such as surface form and grammatically related paraphrasings, by means of a generous variety of transformations: including movement, reattachment, deletion, substitution, copying, lexical insertion, and magic.

The early papers in this collection develop case grammar notions, especially as they figure in the meanings and combinatorial properties of verbs. The later papers, showing a growing interest in the lexicon for its own sake, leak into vaguely delimited areas of pragmatics and language understanding.

ACKNOWLEDGMENTS

Acknowledgment is made to the following publishers and organizations in whose publications the papers in the present book first appeared:

"Toward a Modern Theory of Case" originally appeared in *Modern Studies in English: Readings in Transformational Grammar* (1969), David A. Reibel and Sanford A. Shane, eds., pp. 361–375, Prentice-Hall.

"The Case for Case" originally appeared in *Universals in Linguistic Theory* (1968), Emmon Bach and Robert T. Harms, eds., pp. 1–88, Holt, Rinehart and Winston.

"The Grammar of Hitting and Breaking" originally appeared in *Readings in English Transformational Grammar* (1970), Roderick A. Jacobs and Peter S. Rosenbaum, eds., pp. 120–133, Ginn and Company.

"Types of Lexical Information" originally appeared in *Studies in Syntax and Semantics* (1969), Kluwer Academic Publishers.

"The Case for Case Reopened" originally appeared in *Syntax and Semantics: Grammatical Relations* **8**, 1 (1977), pp. 59–81, Academic Press.

"Topics in Lexical Semantics" originally appeared in *Current Issues in Linguistic Theory* (1977), Roger W. Cole, ed., pp. 76–138, Indiana University Press.

"On the Organization of Semantic Information in the Lexicon" originally appeared in *Papers from the Parasession on the Lexicon, Chicago Linguistic Society, April 14–15, 1978*, Donka Farkas, Wesley M. Jacobsen, Karol W. Todrys, eds., Chicago Linguistic Society.

1

Toward a Modern Theory of Case

Though published in 1969, the writing of this article preceded "The Case for Case", having first appeared in 1966 in an Ohio State University working papers series from Bill Wang's Project on Linguistic Analysis. *The main point of the paper is that the plan of defining grammatical relations in terms of constituent structure configurations cannot yield enough distinctions to cover the kind of clause elements that grammarians need to recognize, and that clause types are best characterized in terms of constellations of semantic role types and their syntactic correlates. The word* case *is not yet used for naming such semantic roles, but Tesnière's (1957) terms* actant *and* circonstant *are mentioned in distinguishing the clausal elements that fit most tightly to a verb's meaning from adverbial modifiers such as indicators of manner, means, purpose, time and place, etc. Putting the core elements in what I called Proposition and the peripheral ones in Modality, I followed the current generative grammar practice of almost completely neglecting the adverbials.*

The key ideas are these: (1) the deepest level of clause structure consist of order-free clusters of elements defined in terms of semantic roles, these represented as named categories (Agent, etc.) that dominated NPs; (2) each NP is provided with a preposition, and these determined sometimes by the lexical verb, sometimes by the dominating semantic role, sometimes by virtue of their independent meanings; (3) transformations operating on these structures yield the familiar ordered kinds of deep structures with both dominance and precedence relations; (4) and, importantly, such transformations include the means of deleting pre-assigned prepositions and placing the resulting noun phrases in clausal positions corresponding to subject or object.

In my teaching I used a hanging mobile to represent the underlying structure; I expressed the hope that such structures could be thought of as language-independent; and I tried to demonstrate that different languages—and different verbs in the same language—provided different ways of flattening the mobile against the wall to get the kinds of structures upon which the "regular" transformations operated. (To support this, I was occasionally able to find or concoct Japanese sentences that could, ignoring the subject, be translated backwards, explaining that this is why Japanese is written from right to left. I stopped this when I found out that some students thought I was serious.)

1.

In Chomsky (1965a, pp. 63–74) we learn of the essentially *relational* nature of such grammatical concepts as subject (of a sentence) and object (of a verb, or of a predicate phrase) as opposed to the *categorial* nature of such notions as verb or noun phrase. The important distinction is there drawn between grammatical relations or grammatical functions, on the one hand, and grammatical categories on the other hand.

The distinction can be captured in formal grammars, according to Chomsky, by introducing category symbols as constituent labels in the phrase structure rules of the base component, and by defining the grammatical relations as in fact relations among category symbols within the underlying phrase-markers provided by the base. Thus sentence, noun phrase, and verb phrase, for example, are provided as category symbols by the base, while the notion subject is defined as a relation between a noun phrase and an immediately dominating sentence, the term object as a relation between a noun phrase and an immediately dominating verb phrase.

My purpose in this essay is to question the deep-structure validity of the notions subject and object, and also to raise doubts about the adequacy of Chomsky's proposals for formally reconstructing the distinction between grammatical categories and grammatical functions. My inquiry will lead to a proposal which renders unnecessary the distinction in English grammar between noun phrase and preposition phrase, and to the suggestion that something very much like grammatical *case* plays a role in the groundwork of grammars that is much less superficial than is usually recognized.

I begin my argument by asking, concerning such expressions as *in the room, toward the moon, on the next day, in a careless way, with a sharp knife,* and *by my brother,* how it is possible in grammars of the type illustrated in *Aspects* to reveal both the categorial information that all of these expressions are preposition phrases and the functional information that they are adverbials of location, direction, time, manner, instrument, and agent respectively. Instead of having a category label *Time,* it ought to be possible–if Chomsky's proposal is adequate–to recognize that a preposition phrase whose head is a

Time noun has the syntactic function *Time Adverbial* within the constituent which immediately contains it.

It seems impossible to provide both types of information in a natural way for the reason that there may be several adverbial expressions in a simple sentence, there are ordering restrictions among these, and if they all start out with the same category, Preposition Phrase, there is no known device by which the further expansion of this category can be constrained according to the permitted order of adverbial types in a single sentence.

Most of the sample phrase structure rules for English that I have seen recently have introduced categorially such terms as *Manner*, *Frequency*, *Extent*, *Location*, *Direction*, etc. In these grammars, for the constituents mentioned, either the strictly categorial information is lost, or else it is rescued by having nonbranching rules which rewrite each of these adverbial-type categories as *Preposition Phrase.* In any case the formal distinction between relations and categories is lost, and the constraints on the further expansion of these preposition phrases that depend on the types of adverbials they manifest need to be provided, as suggested above, in ways that have not yet been made clear.[1]

Other grammars that I have seen contain rules allowing more than one preposition phrase in the expansion of a single category. In the abbreviated form of these rules, each of these preposition phrases is independently optional. Difficulties in establishing the constraints on expanding these categories just in case more than one was chosen remain as before, and two new technical difficulties arise. If there are two independently optional preposition phrases in the expansion of *Verb Phrase*, then we get the same result by skipping the first and choosing the second as we do by choosing the first and skipping the second. The first technical difficulty, then, is that different

---

1. The problem on such restricted expansion has not been ignored. Chomsky (1965a, p. 215) has proposed that "there is some reason to suspect that it might be appropriate to intersperse certain local transformations among the rewriting rules of the base. Thus adverbial phrases consisting of Preposition-Determiner-Noun are in general restricted as to the choice of these elements, and these restrictions could be stated by local transformations to the effect that *Preposition* and *Noun* can be rewritten in certain restricted ways when dominated by such category symbols as *Place Adverbial* and *Time Adverbial.* In fact, one might consider a new extension of the theory of context-free grammar, permitting rules that restrict rewriting by local transformations (i.e., in terms of the dominating category symbol), alongside of the fairly widely studied extension of context-free grammar to context-sensitive grammars that permit rules that restrict rewriting in terms of contiguous symbols." The proposal given below amounts to incorporating Chomsky's suggestions in the form of a convention on the rules which assign complex symbols to prepositions and nouns.

choices in the base do not correspond to differences in the structure of sentences.[2] The second is that now the syntactic relation *preposition-phrase-under-verb-phrase* is not unique in a verb phrase just in case more than one preposition phrase has been chosen.

The obvious alternative within the present conception of grammar is to introduce new structure in such a way that whenever a sentence contains more than one preposition phrase, they are all under immediate domination of categories of different types. If the number of distinct types of preposition phrases is large, this solution differs from providing separate category labels for each adverbial only by greatly increasing the constituent-structure complexity of sentences.

With these difficulties understood, I should next like to ask whether two of the grammatical functions which Chomsky accepts–namely subject and object–are in fact linguistically significant notions on the deep structure level. The deep structure relevance of syntactic functions is with respect to the projection rules of the semantic theory. The semantic component recognizes semantic features associated with lexical elements in a string and projects from them the meaning of the string in ways appropriate to the syntactic relations which hold among these elements. It is my opinion that the traditional subject and object are not to be found among the syntactic functions to which semantic rules must be sensitive.

Consider uses of the verb *open*. It seems to me that in sentences 1 and 2

(1) The door will open.
(2) The janitor will open the door.

there is a semantically relevant relation between *the door* and *open* that is the same in the two sentences, in spite of the fact that the door is the subject of the so-called intransitive verb and the object of the so-called transitive verb. It seems to me, too, that in sentences 3 and 4

(3) The janitor will open the door with this key.
(4) This key will open the door.

the common semantically relevant relation is that between *this key* and *open* in both of the sentences, in spite of the fact that *this key* superficially is the

2. This problem vanishes if parentheses in phrase structure rules are understood as having purely abbreviatory functions.

subject of one of the sentences, the object of a preposition in the other.

In naming the functions of the nominals in these sentences, that of *the janitor* we might call *Agentive*; and that of *this key, Instrumental*. The remaining function to find a name for is that of the subject of an intransitive verb and the object of a transitive verb: a term we might use for this function is *Objective*.[3] None of these functions, as we have seen, can be identified with either subject or object.

If we allow ourselves to use these terms *Objective, Instrumental*, and *Agentive*, we might describe the syntax of the verb *open* as follows: it requires an Objective, and tolerates an Instrumental and/or an Agentive. If only the Objective occurs, the Objective noun is automatically the subject. If an Instrumental also occurs, either the Objective or the Instrumental noun may be the subject, as seen in sentences 5 and 6.

(5) This key will open the door.
(6) The door will open with this key.

If an Agentive occurs, an Instrumental noun cannot be the subject, but, if it occurs, it must appear in a preposition phrase after the Objective, as in 7.

(7) The janitor will open the door with this key.

The Objective noun can be made subject even if the sentence contains Instrumental and Agentive elements, just in case the verb is capable of assuming its passive form. The Instrumental and Agentive expressions, in this case, contain their appropriate prepositions, as in 8 and 9.

(8) The door will be opened with this key.
(9) The door will be opened by the janitor.

In the case of two syntactic functions–Instrumental and Agentive–the noun phrase is preceded by a preposition just in case it has not been made the subject of the sentence. When we add to our consideration the many cases

---

3. The term *Objective* should not be confused with the surface syntactic relation *object* nor with the surface *case accusative*. It would be possible to borrow the term *nominative* from the grammatical tradition of studies of the so-called "ergative" languages (e.g. the northern Caucasian languages), but it is my belief that the choice of this term may well have been responsible for the common assumption that the "ergative" languages are languages in which transitive sentences are obligatorily passive.

where object nouns are also marked by prepositions as in such sentences as 10 and 11

(10) She objects to me.
(11) She depends on me.

and when, further, we see that even in cases like *open*, the Objective has a preposition associated with it in certain nominalizations, as in 12

(12) The opening of the door by the janitor with this key

we see that an analysis of syntactic functions in English requires a general account of the role of prepositions in our language.

The verb *open*, fortunately, is not unique in governing syntactic relations that are not identifiable with subjects and objects. Other verbs that behave in similar ways are *advance, bend, bounce, break, burn up, burst, circulate, close, connect, continue, crumple, dash, decrease, develop, drop; end, enter (contest), expand, hang, hide, hurt, improve, increase, jerk, keep away, keep out, move, pour, repeat, retreat, rotate, run, rush, shake, shift, shine, shrink, sink, slide, spill, spread, stand, start, starve, stir, stretch, turn, twist, wake up, wind, withdraw.* My interpretation of these words is that they have a certain amount of freedom with respect to the syntactic environments into which they can be inserted–a freedom which I assume can be stated very simply. The alternative is to regard these verbs as having each two or three meanings corresponding to their intransitive use or their capability of taking subjects whose relation to the verb can be construed instrumentally in one meaning, agentively in another.[4]

2.

I recognize, therefore, various categorially introduced noun phrase types–suggestive, it seems to me, of the traditional notion of "cases"–each, in English, beginning with a preposition. The syntactic relationship of each of

4. Edward S. Klima has pointed out to me that the instrumental versus agentive role of the possible subjects of *open* has syntactic as well as semantic importance, as can be seen in the unacceptability of the conjunction seen in i:

(i) The janitor and the key will open the door.

Two agentive subjects may be conjoined, two instrumental subjects may be conjoined, but not one of each.

these types to the main verb of the sentence is defined with reference to the category under which it is introduced, having no direct connection with whatever eventual status it may have as subject or object. The following assumptions are meant to develop the scheme I have in mind by means of a series of specific assumptions.

2.1.

The first of these assumptions is that the major constituents of a sentence (S) are Modality (Mod), Auxiliary (Aux), and Proposition (Prop). The first phrase structure rule is 13.

(13) S → Mod - Aux - Prop

I use *Proposition* rather than *Predicate* because it includes what will end up to be the subject of the sentence. Notice that the Auxiliary is in immediate constituent relationship with the entire Proposition, not a subconstituent of the Proposition. I assume this structure assignment to be semantically justified.

2.2.

The constituent Modality contains interrogative and negative elements, sentence adverbials, time adverbials, and various other adverbial elements that are understood as modalities on the sentence as a whole rather than subconstituents of the constituent containing the main verb. I have no strong convictions that these various elements actually comprise a single constituent, but for the time being we may assume that they do. For the purpose of the present discussion I shall also assume that this modality element is optional, and that it is not involved in any of the observations that I shall be dealing with here.[5]

In the rest of the present discussion, however, I shall omit mention of the Modality constituent, acting as if the first rule rewrites *Sentence* as *Auxiliary* + *Proposition.*

5. For various reasons, I am convinced that instead of treating *Negative* as an optional subconstituent of the optional constituent Modality, it is better to introduce, as an obligatory subconstituent of the obligatory constituent Modality, a disjunction of the elements *Negative* and *Affirmative.* This choice appears to be necessary because of various semantic rules whose effect is to reverse the negativity value of a sentence–changing affirmatives to negatives and negatives to affirmatives.

2.3.

The category Proposition includes the verb and all those nominal elements which are relevant to the sub-classification of verbs. The rules for rewriting Proposition take it into an obligatory verb followed by the somewhat independently optional elements Objective (Obj), Dative (Dat), Locative (Loc), Comitative (Com), Instrumental (Ins), and Agentive (Ag).[6]

Roughly speaking, all adverbial elements capable of becoming subjects or objects are introduced in the expansion of Proposition; all others–Time, Benefactive, Frequentative, etc.–are modality elements. The concepts involved are presumed to be among the substantive universals specified by the grammatical theory.

---

6. The variety of expansions of the proposition displays the range of kernel sentence types in the language. Some possible expansions of *Prop* are these (linked parentheses indicate that at least one of two adjacent terms must be chosen):

Prop → V Obj (Dat)(Ag)
Prop → V Obj Loc (Dat)(Ag)
Prop → V Obj (Ins)(Ag)
Prop → V Obj Com
Prop → V S (Dat (Ag))
Prop → V S Obj (Ag)

Sentences illustrating various choices in the expansion of *Prop* are given below:

(i)	John has a car.	V⁀Obj⁀Dat
(ii)	I gave John a car.	V⁀Obj⁀Dat⁀Ag
(iii)	I bought a car.	V⁀Obj⁀Ag
(iv)	A coat is in the closet.	V⁀Obj⁀Loc
(v)	John has a coat in the closet.	V⁀Obj⁀Loc⁀Dat
(vi)	John put a coat in the closet.	V⁀Obj⁀Loc⁀Ag
(vii)	The door opened.	V⁀Obj
(viii)	The key opened the door.	V⁀Obj⁀Ins
(ix)	The janitor opened the door.	V⁀Obj⁀Ag
(x)	The janitor opened the door with the key.	V⁀Obj⁀Ins⁀Ag
(xi)	John is with his brother.	V⁀Obj⁀Com
(xii)	John turned out to be a liar.	V⁀S
(xiii)	John thinks that he is too old.	V⁀S⁀Dat
(xiv)	I persuaded John that he was too old.	V⁀S⁀Dat⁀Ag
(xv)	I forced John to go.	V⁀S⁀Obj⁀Ag

2.4.

All of the nonverb constituents of propositions are Noun Phrases (NP). The relevant rule takes each of the terms I have mentioned and rewrites it as noun phrase:

(14)

$$\begin{Bmatrix} \text{Obj} \\ \text{Dat} \\ \text{Loc} \\ \ldots \end{Bmatrix} \rightarrow \text{NP}$$

The syntactic functions appropriate to noun phrases, in other words, are identified categorially. The elements which dominate NP are distinct from what we might wish to call the true grammatical categories in that their further expansion is unary and many-to-one. What this suggests is that the form of grammars which I am proposing is at bottom one in which the underlying structures of sentences are representable as rooted trees with labeled nodes and labeled branches. This could be equivalent to a phrase structure grammar in which, beginning from S, all even-numbered branchings are unary.

Borrowing from Tesnière (1959, esp. pp. 105–115), I shall use the term *actant* for these elements within propositions which unarily dominate noun phrases.

2.5.

Another important assumption is that every noun phrase begins with a preposition.

(15) NP → P (Det) (S) N

Thus we see that the distinction between noun phrase and preposition phrase is no longer necessary. This is all to the good, of course, since *preposition phrase* has always been a terminological nuisance. We would really like all constituents labeled X-Phrase to be constituents having X's as their heads.

2.6.

The lexical categories Preposition (P) and Noun (N) take by convention the name of the actant dominating their noun phrase as one of the features making up the complex symbols associated with each of these categories. For Agentives, for example, the convention will fill in the feature [+ Agentive] as shown in 16.

(16)

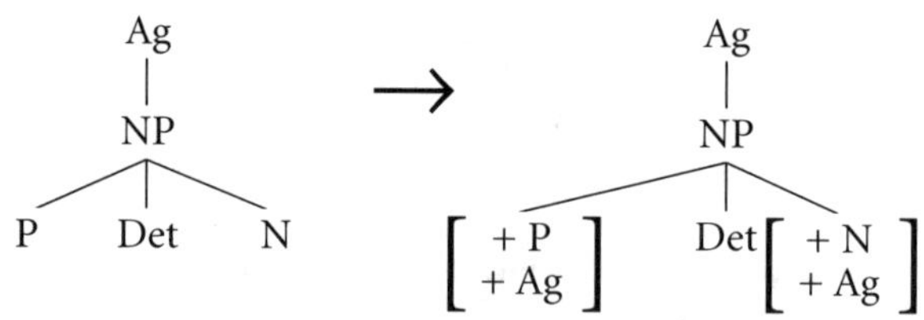

2.7.

The selectional restrictions associated with lexical categories serving given syntactic functions will be provided by appropriate subcategorization. We may wish to guarantee, for instance, that agent nouns are animate, a decision expressed by rule 17.

(17)

$$\begin{bmatrix} +\text{N} \\ +\text{Ag} \end{bmatrix} \rightarrow [+\text{Anim}]$$

The feature [+ Animate] may be required for Benefactives and, under certain environmental conditions, for Datives as well.

2.8.

Some prepositions may be filled in by optional choices from the lexicon. In Locative phrases, though in some cases the preposition may be automatically determined, generally the choice is optional: *over, under, in, on, beside,* etc. These are the prepositions that bring with them semantic information.

2.9.

Other prepositions are determined by inherent syntactic features of specific governing verbs. Thus *blame* requires the Objective preposition to be *for*, the Dative preposition to be *on*.

2.10.

The remaining prepositions are filled in by rules which make use of information about the actants. Thus, the Objective preposition is *of* if it is the only actant in the proposition or if the proposition contains an Instrumental or an Agentive; otherwise it is *with*. The Instrumental preposition is *with* just in case the Agentive co-occurs; otherwise it is *by*.

2.11.

The subject of a sentence is selected, in accordance with certain rules, from among the propositional actants. A transformation places the noun phrase selected to serve as subject to the left of the auxiliary. In Objective-Agentive sentences, unless the auxiliary contains the passive marker, the Agentive becomes the subject, as in 18.[7]

(18)

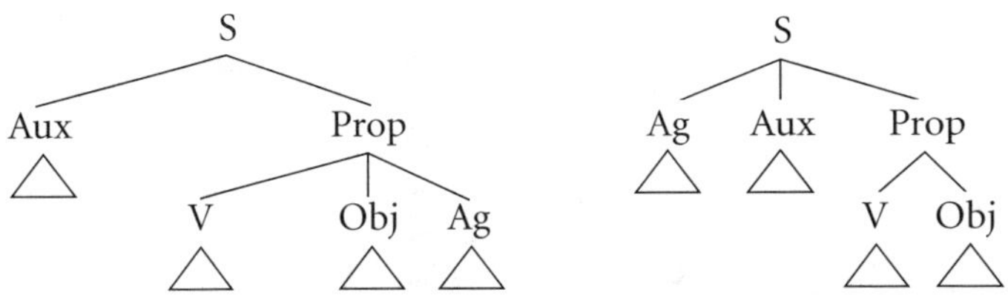

Notice that the proposition that has had one of its actants moved to the subject position is what is traditionally called the predicate.

2.12.

All prepositions are deleted in subject position. Diagram 19 shows that after

7. The triangles represent the subtrees dominated by the categories under which they are drawn.

the Objective *of the dog* becomes the subject of the verb *die*, it loses its preposition.

(19)

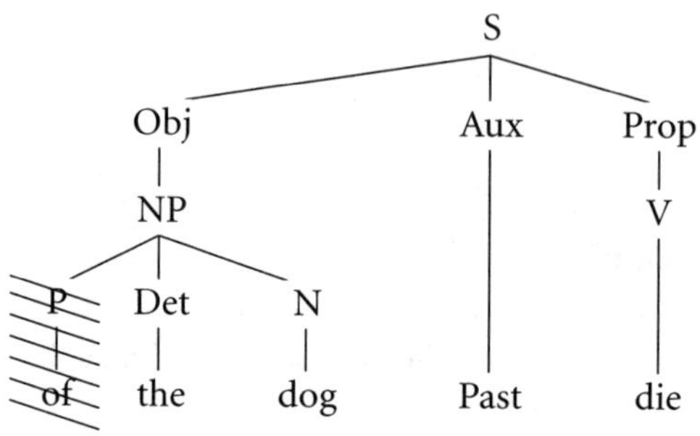

2.13.

A later rule–a rule to which many verbs and certain actants are exceptions—deletes prepositions after verbs:

(20)

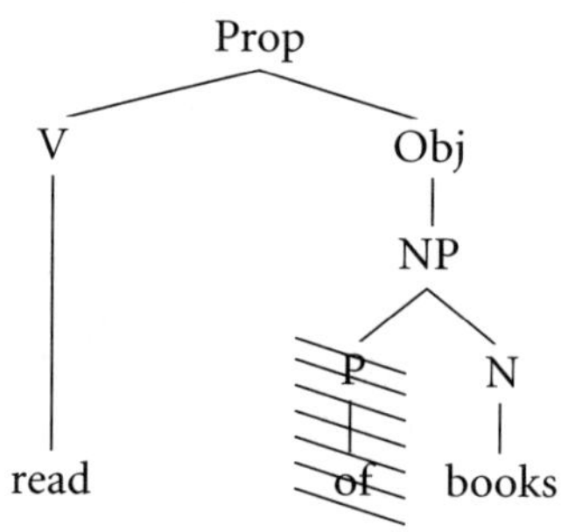

2.14.

Various grammatical processes affect the conditions under which the preposition-deletion rule operates. In nominalizations the Objective preposition stays, as in *the death of the dog* or *the reading of books.*

2.15.

In some contexts, the choice of subject offers certain options. We have seen already that in 21

(21) will open of the door with this key

either the Objective or the Instrumental can be made subject, giving us either 22 or 23.

(22) The door will open with this key.
(23) This key will open the door.

In the sentence whose underlying form is 24

(24) Pres swarm with bees in the garden

either the Objective *with bees* or the Locative *in the garden* may be made subject, losing its preposition in the process. Notice the sentences

(25) Bees swarm in the garden.

where initial *with* has been lost, and

(26) The garden swarms with bees.

where initial *in* has been lost.

2.16.

When Objective and Dative or Objective and Locative are left behind after a subject has been chosen, they may be permuted–subject to certain constraints involving the identification of pronouns. Examples where preposition deletion does not take place are 27 and 28.

(27) Talk about this to Dr. Smith.
(28) Talk to Dr. Smith about this.

Examples with loss of the postverbal preposition are 29

(29) Blame the accident on John.

where *for* was deleted, and 30

(30) Blame John for the accident.

where *on* was deleted. James Heringer of Ohio State University and J. Bruce Fraser of the MITRE Corporation have given me many more examples like this one.[8]

2.17.

In some cases the transformation which provides the subject of a sentence must be thought of as *copying* the selected actant in front of the auxiliary. In Objective-Locative sentences in which the verb is *be*, one subject-selection possibility is the Objective. Thus from 31

(31) Pres be with some books on the shelf.

with some books can be made subject giving us, after the *with* drops out, 32.

(32) Some books are on the shelf.

8. Fraser speaks of these verbs as having alternate meanings. *Spray*, according to Fraser, has one meaning in i

(i) Spray the wall with paint.

another in ii.

(ii) Spray paint on the wall.

I would say merely that the nonagentive actants associated with *spray* are, in these sentences, *with paint* and *on the wall.* They may occur in either order, but whichever one comes first loses its preposition. Other examples from Fraser are iii to x:

(iii) Stuff cotton into the sack.
(iv) Stuff the sack with cotton.
(v) Plant the garden with roses.
(vi) Plant roses in the garden.
(vii) Stack the table with dishes.
(viii) Stack dishes onto the table.
(ix) Make a chair out of wood.
(x) Make wood into a chair.

Alternatively, however, the Locative actant may be *copied* in the subject position, giving us 33.

(33) on the shelf are with some books on the shelf

If the left copy of the Locative actant is replaced by its pro-form, we end up with 34.

(34) There are some books on the shelf.[9]

2.18.

The verb *have* is here interpreted as a variant of the verb *be* in front of the Objective after a noun phrase. Whenever a Locative or a Comitative is made the subject of a proposition whose verb is *be*, this is done by subject copying. Unless, as in the existential sentences, the left copy is changed to *there*, the nominal part of the right copy will be pronominalized. (Possibly what this means is that the repeated noun gets deleted and the features that at this stage have been assigned to the determiner serve now to select a pronoun.)

The Locative copying can be seen in 35:

---

9. I would expect that the expletive *it* can be handled in an analogous way. Nominalized sentences are copied in the subject position, giving us something like i.

(i) that he is a liar is true that he is a liar

Now, either the right copy can be deleted or the left copy can be replaced by *it*, resulting in ii or iii, respectively.

(ii) That he is a liar is true.
(iii) It is true that he is a liar.

It is likely that the *copying* method of providing subjects should be generalized to all cases, with, simply, the right copy getting deleted in a majority of cases. Where the expected deletion of the right copy is not effected, we get such somewhat deviant sentences as iv and v, borrowed from Jespersen,

(iv) He is a great scoundrel, that husband of hers.
(v) It is perfectly wonderful the way in which he remembers things.

because where the right copy didn't get deleted, the left copy got pronominalized. This seems, too, to be a way of handling cases where a preposition gets left behind in passive sentences. The deletion of the right copy is sometimes only partial, leaving the preposition behind, as in

(vi) Mary can be depended on.

(35)

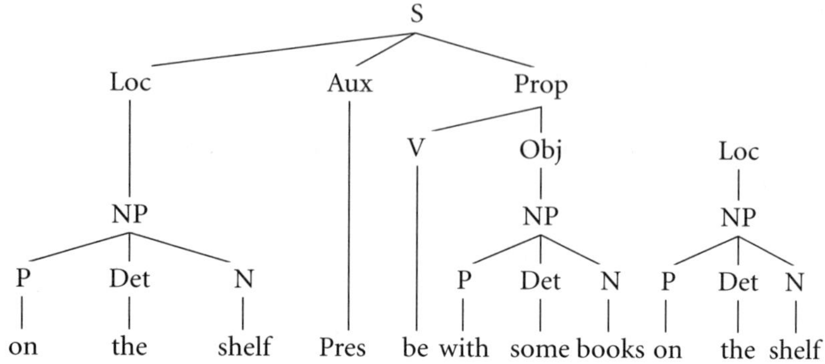

From 35 we get, at a certain stage, something like 36.

(36) *on the shelf is with some books on the shelf

The sentence-initial preposition is deleted, the repeated phrase *the shelf* is replaced by *it*, *with* is deleted, and *be*, after noun phrase and before an Objective, gets changed to *have*[10]. The resulting sentence is 37

(37) The shelf has some books on it.

which, thus, I regard as a simplex.

Analogous phenomena are observed in Objective-Comitative sentences. There too we find *have* only when the Objective is not the subject. In sentence 38

(38) The children are with Mary.

the Objective is the subject, *be* remains unchanged, and the Comitative preposition stays. If the Comitative is made subject, it is done so by copying, the right copy eventually undergoing pronominalization. Here *be* is followed by

10. In my present view of the design of a grammar I hold that the lexicon may be divided into two sections, a major-category lexicon which inserts semantically relevant lexical items into underlying phrase-markers, and a minor-category lexicon which inserts "function words" into surface phrase-markers. If *be* and *have* are regarded as words in the minor-category lexicon, it will not be necessary to speak, as I have here, of "changing *be* to *have*." The structural conditions for inserting *be* and *have* will simply be different.

an Objective, so it becomes *have*, and in the Objective case the preposition is deleted. Whereas before we had 38 with *are*, we now have 39 with *has*.

(39) Mary has the children with her.

In Objective-Dative sentences, at least for those cases where the Dative noun is animate, the Dative becomes subject by transposition rather than by copying. (Alternatively we can say that in this case the right copy is always deleted).

With Objective-Dative sentences it appears that the choice of subject is determined by the verb. The verb *belong*, for example, requires the Objective to be subject, as in 40

(40) The typewriter belongs to Terry.

while the verb *be* requires the Dative as subject, as in 41.

(41) Terry has the typewriter.

2.19.

Verbs in the lexicon will be classified according to the propositional environments into which they may be inserted. Using brute force methods for the time being I allow options in the statement of these environments. Thus a verb like *wake up* would have the feature 42

(42) [+ ____ Obj (Inst) (Ag)]

while a verb like *kill* would have the feature 43

(43) [+ ____ Obj (Inst)(Ag)]

where the linked parentheses indicate that at least one of two adjacent terms must be chosen.

For *wake up* we have sentences like 44 with Objective only; 45 with Agentive; 46 with Instrumental; and 47 with both Agentive and Instrumental in addition to the Objective.

(44) I woke up.

(45) My daughter woke me up.
(46) An explosion woke me up.
(47) My daughter woke me up with an explosion.

For *kill*, on the other hand, we have 48–50 but not 51.

(48) Fire killed the rats.
(49) Mother killed the rats.
(50) Mother killed the rats with fire.
(51) *The rats killed.

2.20.

Notions of synonymy can now be separated from notions of syntactic distribution. The verbs *kill* and *die*, for instance, may be given the same semantic characterizations. The relation between the verb and the Objective is the same for both words, the difference between the verbs being of a syntactic nature, in that *kill* requires a co-occurring Agent or Instrument, *die* does not allow an Agent to be directly expressed as part of the same proposition. In this respect, the essential difference between the proposals presented in this paper and those of Lakoff (1965) appears to be that Lakoff seeks for "synonymous" words identity of semantic reading and what he calls the *lexical base*, but not identity in *lexical extension*; I seek only identity of semantic reading. In other words, I do not expect to find in formal grammars support for the distributional definition of meaning.

3.

I believe that certain advantages derive from incorporating into a transformational grammar the proposals that have just been sketched, in addition to the possibly unimportant one that sentences do not turn out to need quite so much branching structure as they do in grammars that need to recognize syntactic relations in terms of immediate-domination relation between categories.

3.1.

One of the specific advantages of my interpretation of *have* is in the simplification this analysis allows to the relative clause reduction rule.

I have said that the Objective preposition in sentences with Datives, Locatives, or Comitatives but without Instrumentals and Agentives is *with*. The preposition appears in this form in a sentence like 26, but it disappears after *be* under those conditions which ordinarily change *be* to *have*, as in 52.

(52) The garden has bees in it.

In older versions of the grammar of English, one relative clause reduction rule was needed for relative-clause-plus-*be*, changing 53 into 54

(53) the boy who is in the next room
(54) the boy in the next room

and another rule was needed for relative-pronoun-plus-*have*, changing 55 into 56.

(55) the boy who has the red hat
(56) the boy with the red hat

The first of these rules deleted the identified element, the second replaced the identified element by *with*.

If it is true, however, that *have*, abstractly, is *be* before Objectives, then a single rule will now cover both of these cases. From 53 we get 54 and from 57 we get 56.

(57) the boy who is with the red hat

We need to require only that the relative-clause reduction rule precede the rules for creating *have* and deleting *with* after *be*.[11]

3.2.

Notions like that of the "understood agent" can be clarified within this scheme. There is a distinction, in other words, between not choosing an Agent on the one hand, and choosing an Agent and subsequently deleting it on the other hand. The distinction is revealed in the choice of preposition.

The verb *kill*, I have said, must take either Instrumental or Agentive and

11. But see footnote 10.

may take both. The Instrumental preposition is *by* if there is no Agentive present, otherwise it is *with.* The Agentive–as in the case of passive sentences–may be a dummy.

Suppose that we construct passive sentences with *kill* where the Objective is *the rats* and the Instrumental is *fire.* In one case we will omit an Agentive, in the other case the Agentive will be chosen but it will be a dummy. Where the Agentive is present in the deep structure, the Instrumental preposition is *with*; where there is no Agentive, the Instrumental preposition is *by.* Since the Agentive is a dummy, it gets deleted.

The resulting sentences are 58 and 59.

(58) The rats were killed by fire.
(59) The rats were killed with fire.

If my analysis is correct, there is an "understood agent" in the sentence with *with.*

Incidentally, the earlier examples with *open* were a little misleading. I implied that in a sentence like 60

(60) The door was opened with this key.

the underlying representation of the sentence contained only the actants Objective and Instrumental. If the above observations on *with* and *by* are true, however, the sentence should actually be understood as having an implied human agent, and it should be distinct in this respect from a sentence like 61; and I believe it is.

(61) The door was opened by the wind.

3.3.

More general advantages associated with these proposals relate to the interpretation of historical changes and cross-language differences in lexical structure.

3.3.1. Certain historical changes in language may turn out to be purely syntactic, and, in fact, may pertain exclusively to the status of particular lexical items as exceptions to given transformations, in the sense of Lakoff. Thus the English verb *like* did not change in its meaning or in its selection for Objective-Dative sentences, but only in its status as an exception to the rule that fronted actants are neutralized to the so-called nominative form.

3.3.2. Lexical differences across languages may not be as great as we might otherwise have thought. It would ordinarily be said that English *kill* and Japanese *korosu* have different "meanings" because the Japanese verb requires an animate subject while English allows us to say that *fire killed him, a falling stone killed him*, and the like. Once we see that even in English both *kill* and *die* have the same underlying semantic representation, the difference between the two situations appears to be rather superficial. Both languages have words with the same meaning which can co-occur with Objective and Instrumental. English has two such verbs, one of which allows the Instrumental phrase to become the subject. The difference is no deeper than that.

4.

There are, as the reader may have guessed, a great many extremely serious problems which continue to be completely mysterious. Does this system provide the constituent structure needed for co-ordinate conjunction? How are predicate-adjective or predicate-noun sentences to be dealt with in this scheme? Do manner adverbials belong inside the proposition or are they part of the modality? How is the relation sometimes found between manner adverbials and the "subject" of the sentence to be expressed in this system? What about the generalizations on noun phrase interchange? Many of these problems, fortunately, are no *less* serious in subject-object grammars.

I could summarize my remarks by saying that I regard each simple sentence in a language as made up of a verb and a collection of nouns in various "cases" in the deep structure sense. In the surface structure, case distinctions are sometimes preserved, sometimes not—depending on the language, depending on the noun, or depending on idiosyncratic properties of certain governing words. Belief in the superficiality of grammatical case rises from consideration of the "nominative," which really constitutes a case neutralization that affects noun phrases that have been made the subject of a sentence, and of the "genitive," which represents another kind of neutralization of case distinctions, one which occurs in noun phrase modifiers derived from sentences, as illustrated by the reduction to the so-called "genitive" of both Agentive and Objective in such expressions as *the shooting of the hunters.*

2

The Case for Case

This long paper was written at OSU during a quarter's release from teaching in 1967. My sabbatical was granted on the promise that I would write something on deixis, but the invitation from the Texas conference on language universals distracted me and I produced this instead. The long list of influences on this paper includes: traditional grammarians' descriptions of the classical "case languages" in which names referring to semantic functions were considered appropriate for the "oblique" cases, but not for the "straight" ones (nominative and accusative); an old LSA monograph by Blake in which the word case *was proposed in very much the sense that I needed, setting it off from* case form, *which covers the morphologically marked categories; the numerous discussions of the kinds of paraphrase relations (now spoken of as valence variation) of the kind that showed syntactic rearrangements around the same lexical verb; the generous use of "optional" functions in the tagmemic formulas of Kenneth Pike and his followers; the* petit drame *image of the semantic structure of a clause in the work of Lucien Tesnière; the precursors to Generative Semantics in the writings of George Lakoff and Paul Postal which welcomed a way of thinking in which sentences that had "the same meaning" also had the same underlying structure, and transformations of a variety of types were invented to make things come out right.*

The case grammar contribution to generative grammar was thought to be one of providing a kind of "deeper deep structure", common to all sentences that had the same meaning–in those days we referred to the common bit as the sentence's "cognitive meaning" as opposed to whatever contribution was made by the surface syntactic organization–and specifying the transformations that could convert these to something like the familiar Chomsky-standard Deep Structures upon which the "regular" transformations of the language would operate. (Passive and Dative, I saw as operating in the pre-Deep Structure series, leaving the various clean-up transformations among the "regulars"). In the end–for exam-

Originally published in 1968

ple, in the closing words of this article–I came to assume that the intervening "standard" Deep Structure level was not needed.

A case hierarchy was proposed to guide the mapping from the semantic cases to the subject and object functions; passive verb morphology was interpreted as a way of "registering" a non-canonical subject choice. The recognition of paraphrase relations was seen as a reason for inventing identical "case bases" with transformations leading to the surface realizations; thus sentences like "toys are in the box", "there are toys in the box" and "the box has toys in it" start out with a single semantic relation between the toys and the box and the transformations upon that single underlying pattern provide alternative ways of expressing the notions.

Several proposals were offered here for giving a case-grammar basis for a number of claims in the area of language typology.

Speculation on language universals has not always and everywhere been viewed as a fully respectable pastime for the scientific linguist. The writer recalls a Linguistic Institute lecture of not many summers ago in which it was announced that the only really secure generalization on language that linguists are prepared to make is that 'some members of some human communities have been observed to interact by means of vocal noises'. Times have changed, it is a pleasure to report, and this is partly because we now have clearer ideas about what linguistic theories are theories of, and partly because some linguists are willing to risk the danger of being dead wrong.[1]

Scholars who have striven to uncover syntactic features common to all of the world's languages have generally addressed themselves to three intimately related but distinguishable orders of questions: (*a*) What are the formal and substantive universals of syntactic structure? (*b*) Is there a universal base, and, if so, what are its properties? (*c*) Are there any universally valid constraints on the ways in which deep structure representations of sentences are given expression in the surface structure?

Concerning formal universals we find such proposals as Chomsky's, that each grammar has a base component capable of characterizing the underlying syntactic structure of just the sentences in the language at hand and containing at least a set of transformation rules whose function is to map the

1. I am grateful to the College of Arts and Sciences of the Ohio State University for releasing me from teaching duties during the winter quarter of 1967. I wish also to express my appreciation to George Lakoff (Harvard), D. Terence Langendoen (Ohio State), and Paul M. Postal (I.B.M.) for the many challenges and suggestions they have sent my way concerning the ideas of this paper. I may soon regret that I did not always follow their advice.

underlying structures provided by the base component into structures more closely identifiable with phonetic descriptions of utterances in that language (Chomsky 1965, pp. 27–30). A representative statement on substantive syntactic universals is Lyons' assertion (1966, pp. 211, 223) that every grammar requires such categories as Noun, Predicator, and Sentence, but that other grammatical categories and features may be differently arranged in different languages. And Bach (1965) has given reasons to believe that there is a universal set of transformations which each language draws from in its own way, and he has shown what such transformations might look like in the case of relative clause modification.

Discussions on the possibility of a universal base (as distinct from claims about universal constraints on the form of the base component) have mainly been concerned with whether the elements specified in the rules of a universal base–if there is one–are sequential or not. A common assumption is that the universal base specifies the needed syntactic relations, but the assignment of sequential order to the constituents of base structures is language specific. Appeals for sequence-free representations of the universal deep structure have been made by Halliday (1966), Tesnière (1959), and others. Lyons (1966, p. 227) recommends leaving for empirical investigation the question of the relationship between the underlying representation and sequential order, and Bach (1965) has suggested that continued investigation of the syntactic rules of the world's languages may eventually provide reasons for assuming specific ordering relations in the rules of a universal base.

Greenberg's (1963) statistical studies of sequence patterns in selected groups of languages do not, it seems to me, shed any direct light on the issue at hand. They may be regarded as providing data which, when accompanied by an understanding of the nature of syntactic processes in the specific languages, may eventually lend comfort to some proposal or other on either the sequential properties of the base component or the universal constraints which govern the surface ordering of syntactically organized objects.

Findings which may be interpreted as suggesting answers to our third question are found in the 'markedness' studies of Greenberg (1966) and in the so-called implicational universals of Jakobson (1958a). If such studies can be interpreted as making empirical assertions about the mapping of deep structures into surface structures, they may point to universal constraints of the following form: While the grammatical feature 'dual' is made use of in one way or another in all languages, only those languages which have some overt morpheme indicating 'plural' will have overt morphemes indicating 'dual'. The theory of implicational universals does not need to be interpreted, in other words, as a set of assertions on the character of possible deep structures in human languages and the ways in which they differ from one another.

The present essay is intended as a contribution to the study of formal and substantive syntactic universals. Questions of linear ordering are left untouched, or at least unresolved, and questions of markedness are viewed as presupposing structures having properties of the kind to be developed in these pages.

My paper will plead that the grammatical notion 'case' deserves a place in the base component of the grammar of every language. In the past, research on 'case' has amounted to an examination of the variety of semantic relationships which can hold between nouns and other portions of sentences; it has been considered equivalent to the study of semantic functions of inflectional affixes on nouns or the formal dependency relations which hold between specific nominal affixes and lexical-grammatical properties of neighboring elements; or it has been reduced to a statement of the morphophonemic reflexes of a set of underlying 'syntactic relations' which themselves are conceived independently of the notion of 'case'. I shall argue that valid insights on case relationships are missed in all these studies, and that what is needed is a conception of base structure in which case relationships are primitive terms of the theory[2] and in which such concepts as 'subject' and 'direct object' are missing. The latter are regarded as proper only to the surface structure of some (but possibly not all) languages.

Two assumptions are essential to the development of the argument, assumptions that are, in fact, taken for granted by workers in the generative grammar tradition. The first of these is *the centrality of syntax.* There was a time when a typical linguistic grammar was a long and detailed account of the morphological structure of various classes of words, followed by a two- or three-page appendix called 'Syntax' which offered a handful of rules of thumb on how to 'use' the words described in the preceding sections–how to combine them into sentences.

In grammars where syntax is central, the forms of words are specified with respect to syntactic concepts, not the other way around. The modern grammarian, in other words, will describe the 'comparative construction' of a given language in the most global terms possible, and will then add to that a

2. Notational difficulties make it impossible to introduce 'case' as a true primitive as long as the phrase-structure model determines the form of the base rules. My claim is then, that a designated set of case categories is provided for every language, with more or less specific syntactic, lexical, and semantic consequences, and that the attempt to restrict the notion of 'case' to the surface structure must fail.

description of the morphophonemic consequences of choosing particular adjectives or quantifiers within this construction. This is altogether different from first describing the morphology of words like *taller* and *more* and then adding random observations on how these words show up in larger constructions.[3]

The second assumption I wish to make explicit is *the importance of covert categories.* Many recent and not-so-recent studies have convinced us of the relevance of grammatical properties lacking obvious 'morphemic' realizations but having a reality that can be observed on the basis of selectional constraints and transformational possibilities. We are constantly finding that grammatical features found in one language show up in some form or other in other languages as well, if we have the subtlety it takes to discover covert categories. Incidentally, I find it interesting that the concept 'covert category'– a concept which is making it possible to believe that at bottom all languages are essentially alike–was introduced most convincingly in the writings of Whorf, the man whose name is most directly associated with the doctrine that deep-seated structural differences between languages determine the essentially noncomparable ways in which speakers of different languages deal with reality (see Whorf 1965, pp. 69 ff.).

One example of a 'covert' grammatical distinction is the one to which traditional grammarians have attached the labels '*affectum*' and '*effectum*', in German '*affiziertes Objekt*' and '*effiziertes Objekt*'. The distinction, which is reportedly made overt in some languages, can be seen in sentences 1 and 2.

(1) John ruined the table.
(2) John built the table.

Note in one case the object is understood as existing antecedently to John's activities, while in the other case its existence resulted from John's activities.

3. John R. Ross pointed out, during the symposium, that some syntactic processes seem to depend on (and therefore 'follow') particular lexical realizations of just such entities as the comparative forms of adjectives. Compared adjectives, in short, may be iterated, just as long as they have all been given identical surface realizations. One can say,

(i) She became friendlier and friendlier.
(ii) She became more and more friendly.

but not

(iii) *She became friendlier and more friendly.

Having depended so far on only 'introspective evidence', we might be inclined to say that the distinction is purely a semantic one, one which the grammar of English does not force us to deal with. Our ability to give distinct interpretations to the verb–object relation in these two sentences has no connection, we might feel, with a correct description of the specifically syntactical skills of a speaker of English.

The distinction does have syntactic relevance, however. The *effectum* object, for example, does not permit interrogation of the verb with *do to*, while the *affectum* object does. Thus one might relate sentence 1, but not sentence 2, to the question given in 3.

(3) What did John do to the table?

Furthermore, while sentence 1 has sentence 4 as a paraphrase, sentence 5 is not a paraphrase of sentence 2.

(4) What John did to the table was ruin it.
(5) What John did to the table was build it.[4]

To give another example, note that both of the relationships in question may be seen in sentence 6 but that only in one of the two senses is sentence 6 a paraphrase of sentence 7.

(6) John paints nudes.
(7) What John does to nudes is paint them.

There is polysemy in the direct object of 6, true, but the difference also lies in whether the objects John painted existed before or after he did the painting.

I am going to suggest below that there are many semantically relevant syntactic relationships involving nouns and the structures that contain them, that these relationships–like those seen in 1 and 2–are in large part covert but are nevertheless empirically discoverable, that they form a specific finite set, and that observations made about them will turn out to have considerable cross-linguistic validity. I shall refer to these as 'case' relationships.

1. Earlier Approaches to the Study of Case

Books written to introduce students to our discipline seldom fail to acquaint

4. This observation is due to Paul M. Postal.

their readers with the 'wrong' ways of using particular case systems as universal models for language structure. Grammarians who accepted the case system of Latin or Greek as a valid framework for the linguistic expression of all human experience were very likely, we have been told, to spend a long time asking the wrong kinds of questions when they attempted to learn and describe Aleut or Thai. We have probably all enjoyed sneering, with Jespersen, at his favorite 'bad guy', Sonnenschein, who, unable to decide between Latin and Old English, allowed modern English *teach* to be described as either taking a dative and an accusative, because that was the pattern for Old English *tæcan*, or as taking two accusatives, in the manner of Latin *doceo* and German *lehren* (Jespersen 1924a, p. 175).

Looking for one man's case system in another man's language is not, of course, a good example of the study of case. The approaches to the study of case that do need to be taken seriously are of several varieties. Many traditional studies have examined, in somewhat semantic terms, the various *uses* of case. More recent work has been directed toward the analysis of the case *systems* of given languages, under the assumptions suggested by the word 'system'. A great deal of research, early and late, has been devoted to an understanding of the *history* or *evolution* of case notions or of case morphemes. And lastly, the generative grammarians have for the most part viewed case markers as surface structure reflexes, introduced by rules, of various kinds of deep and surface syntactic relations.

1.1. *Case Uses*

The standard handbooks of Greek and Latin typically devote much of their bulk to the classification and illustration of semantically different relationships representable by given case forms. The subheadings of these classifications are most commonly of the form *X of Y*, where *X* is the name of a particular case and *Y* is the name for a particular 'use' of *X*. The reader will recall such terms as 'dative of separation', 'dative of possession', and so on.[5]

Apart from the fact that such studies do not start out from the point of view of the centrality of syntax, the major defects of these studies were (*a*) that the nominative was largely ignored and (*b*) that classificatory criteria which ought to have been kept distinct were often confused.

The neglect of the nominative in studies of case uses probably has several sources, one being the etymological meaning ('deviation') of the Greek term for case, *ptôsis*, which predisposed grammarians to limit the term only to the

5. For an extensive description of this type, see Bennett (1914).

nonnominative cases. The most important reason for omitting the nominative in these studies, however, is the wrongly assumed clarity of the concept 'subject of the sentence'. Müller (1908, p. 1) published a study of nominative and accusative case uses in Latin, in which he devoted 170 or so pages to the accusative and somewhat less than one page to the nominative, explaining that '*die beiden casus recti, der Nominativ und der Vokativ, sind bei dem Streite uber die Kasustheorie nicht beteiligt. Im Nominativ steht das Subjekt, von dem der Satz etwas aussagt*'.

The role of the subject was so clear to Sweet that he claimed that the nominative was the only case where one could speak properly of a 'noun'. He viewed a sentence as a kind of predication on a given noun, and every noun-like element in a sentence other than the subject as a kind of derived adverb, a part of the predication.[6]

On a little reflection, however, it becomes obvious that semantic differences in the relationships between subjects and verbs are of exactly the same order and exhibit the same extent of variety as can be found for the other cases. There is in principle no reason why the traditional studies of case uses fail to contain such classifications as 'nominative of personal agent', 'nominative of patient', 'nominative of beneficiary', 'nominative of affected person', and 'nominative of interested person' (or, possibly, 'ethical nominative') for such sentences as 8 to 12, respectively.

(8) He hit the ball.
(9) He received a blow.
(10) He received a gift.
(11) He loves her.
(12) He has black hair.

The confusion of criteria in treatments of the uses of cases has been documented by de Groot (1956) in his study of the Latin genitive. Uses of cases are classified on syntactic grounds, as illustrated by the division of uses of the genitive according to whether the genitive noun is in construction with a noun, an adjective, or a verb; on historical grounds, as when the uses of the syncretistic Latin ablative case are divided into three classes, separative, locative, and instrumental; and on semantic grounds, in which there is a great deal of confusion between meanings that can properly be thought of as associated with the case forms of nouns, on the one hand, and meanings that

6. Quoted in Jespersen (1924, p. 107).

properly reside in neighboring words.

De Groot's critical treatment of the traditional classification of Latin genitive case uses is particularly interesting from the point of view taken here, because in his 'simplification' of the picture he rejects as irrelevant certain phenomena which generative grammarians would insist definitely *are* of syntactic importance. He claims, for example, that the traditional studies confuse difference of *referents* with differences of case uses. Thus, to de Groot (1956, p. 35) the traditional three senses of *statua Myronis* (the statue possessed by Myro–*genitivus possessivus*; statue sculpted by Myro–*genitivus subjectivus*; statue depicting Myro–genitive of represented subject), as well as the subjective and objective senses of *amor patris*, are differences in practical, not in linguistic, facts. From arguments such as this he is able to combine twelve of the classical 'uses' into one, which he then labels the 'proper genitive', asserting that 'the proper genitive denotes, and consequently can be used to refer to, *any* thing-to-thing relation'. He ends by reducing the thirty traditional 'uses of the genitive' to eight,[7] of which two are rare enough to be left out of consideration, and a third, 'genitive of locality', is really limited to specific place names.

Benveniste (1962) replied to de Groot's analysis in the issue of *Lingua* that was dedicated to de Groot. There he proposes still further simplifications of the classification. Noting that de Groot's 'genitive of locality' applies only to proper place names, that is, that it occurs only with place names having *-o-* and *-ā-* stems, in complementary distribution with the ablative, Benveniste wisely suggests that this is something that should be catalogued as a fact about place names, not as a fact about uses of the genitive case. Benveniste's conclusions on the remaining genitive constructions is quite congenial to the generative grammarian's position. He proposes that the so-called proper gen-

7. From de Groot (1956, p. 30):

- I. adjunct to a noun
  - A. proper genitive, *eloquentia hominis*
  - B. genitive of quality, *homo magnae eloquentiae*
- II. adjunct to a substantival
  - C. genitive of the set of persons, *reliqui peditum*
- III. conjunct ('complement') of a copula
  - D. genitive of the type of person, *sapientis est aperte odisse*
- IV. adjunct to a verb
  - E. genitive of purpose, *Aegyptum profiscitue cognoscende antiquitatis*
  - F. genitive of locality, *Romae consules creabantur*
- IVa. adjunct to a present participle
  - G. genitive with a present participle, *laboris fugiens*
- V. genitive of exclamation, *mercimoni lepidi*

itive basically results from the process of converting a sentence into a nominal. The distinction of meaning between 'genitivus subjectivus' and 'genitivus objectivus' constructions merely reflects the difference between situations in which the genitive noun is an original subject and those where it is an original object, the genitive representing a kind of neutralization of the nominative/accusative distinction found in the underlying sentences.[8]

At least from the two mentioned studies of uses of the Latin genitive, it would appear (*a*) that some case uses are purely irregular, requiring as their explanation a statement of the idiosyncratic grammatical requirements of specific lexical items, and (*b*) that some semantic differences are accounted for independently of assigning 'meanings' to cases, either by recognizing meaning differences in 'governing' words or by noting meaning differences in different underlying sentences. The suggestion that one can find clear special meanings associated with surface cases fails to receive strong support from these studies.

1.2. *Case Systems*

There are reasonable objections to approaching the case system of one language from the point of view of the surface case system of another (for example, Classical Latin) by merely checking off the ways in which a given case relation in the chosen standard is given expression in the language under observation. An acceptable alternative, apparently, is the inverse of this process: one identifies case morphemes in the new language within the system of noun inflection and then relates each of these to traditional or 'standard' case notions. To take just one recent example, Redden (1966) finds five case indices in Walapai (four suffixes and zero) and identifies each of these with terms taken from the tradition of case studies: *-c* is nominative, *-Ø* is accusative, *-k* is allative/adessive, *-l* is illative/inessive, and *-m* is ablative/abessive. Under each of these headings the author adds information about those uses of each case form that may not be deducible from the labels themselves. Nominative, for example, occurs only once in a simple sentence–coordinate conjunction

8. It must be said, however, that Benveniste's desentential interpretation is diachronic rather than synchronic, for he goes on to explain that it is an analogy from these basic verbal sources that new genitive relations are created. From *luaus pueri* and *risus pueri*, where the relation to *ludit* and *ridet* is fairly transparent, the pattern was extended to include *somnus pueri*, *mos pueri*, and finally *fiber pueri*. The generative grammarian may be inclined to seek synchronic verbal connections–possibly through positing abstract entities never realized as verbs–for these other genitives too. (See Benveniste 1962, p. 17.)

of subject nouns requires use of the *-m* suffix on all the extra nouns introduced; accusative is used with some noun tokens which would not be considered direct objects in English; allative/adessive has a partitive function; and ablative/abessive combines ablative, instrumental, and comitative functions.

In a study of this type, since what is at hand is the surface structure of the inflection system of Walapai nouns, the descriptive task is to identify the surface case forms that are distinct from each other in the language and to associate 'case functions' with each of these. What needs to be emphasized is (*a*) that such a study does not present directly available answers to such questions as 'How is the indirect object expressed in this language?' (for example, the system of possible case functions is not called on to provide a descriptive framework), and (*b*) that the functions or uses themselves are not taken as primary terms in the description (for example, the various 'functions' of the 'ablative/abessive' suffix *-m* are not interpreted as giving evidence that several distinct cases merely happen to be homophonous).[9]

One approach to the study of case systems, then, is to restrict oneself to a morphological description of nouns and to impose no constraints on the ways in which the case morphemes can be identified with their meanings or functions. This is distinct from studies of case systems which attempt to find a unified meaning for each case. An example of the latter approach is found in the now discredited 'localistic' view of the cases in Indo-European, by which dative is 'the case of rest', accusative 'the case of movement to', and genitive 'the case of movement from'.[10] And recent attempts to capture single comprehensive 'meanings' of the cases have suffered from the vagueness and circularity expected of any attempt to find semantic characterizations of surface-structure phenomena.[11]

The well-known studies of Hjelmslev (1935, 1937) and Jakobson (1936a) are

9. These remarks are not intended to be critical of Redden's study. Indeed, in the absence of a universal theory of case relationships there is no theoretically justified alternative to this approach.

10. This interpretation, discussed briefly in Jespersen (1924a, p. 186), appears to date back to the Byzantine grammarian Maxime Planude.

11. As an illustration of this last point, take Gonda's claim (1962, p. 147) that the Vedic dative is called for whenever a noun is used to refer to the 'object in view'. The vacuity of this statement is seen in his interpretation of

> *vātāya hapilā vidyut* (Patanjali)
> 'a reddish lightning signifies wind'

as 'the lightning has, so to say, *wind in view*'.

attempts not only to uncover unified meanings of each of the cases, but also to show that these meanings themselves form a coherent system by their decomposability into distinctive oppositions. The possibility of vagueness is, of course, increased inasmuch as the number of oppositions is less than the number of cases.[12]

The difficulties in discovering a unified meaning for each of the cases in a case system have led to the alternative view that *all but one* of the cases can be given more or less specific meanings, the meaning of the residual case being left open. This residual case can either have whatever relation to the rest of the sentence is required by the meanings of the neighboring words, or it can serve any purely caselike function not preempted by the other cases. Bennett tells us that Goedicke explained the accusative as 'the case used for those functions not fulfilled by the other cases'. The fact that Bennett, following Whitney, ridiculed this view on the grounds that *any* case could be so described suggests that Goedicke's remark must not have been very clearly expressed.[13] A different approach is taken by Diver (1964), who assigns the 'leftover' function not to a particular case as such, but to whatever case or cases are not required for a given realization of what he calls the 'agency system'. Briefly, and ignoring his treatment of passive sentences, Diver's analysis is this: A verb can have one, two, or three nouns (or noun phrases) associated with it, corresponding generally to the intransitive, normal transitive, and transitive indirect object sentence types, respectively. In a three-noun sentence, the nouns are nominative, dative, and accusative, the nominative being the case of the agent and the accusative the case of the patient; the dative, the 'residue' case, is capable of expressing any notion compatible with the meaning of the remainder of the sentence. The function of the dative in a three-noun sentence, in other words, is 'deduced' from the context; it is not present as one of a number of possible 'meanings' of the dative case.[14] In two-noun sentences, one of the nouns is

12. See, in this regard, the brief critical remarks of A. H. Kuipers (1962, p. 231).

13. Bennett (1914, p. 195, fn. 1). I have not yet had access to the Goedicke original.

14. The following is from Diver, 1964, p. 181:

> In the sentence *senatus imperium mihi dedit* 'the senate gave me supreme power', the Nominative, with the syntactic meaning of Agent, indicates the giver; the Accusative, with the syntactic meaning of Patient, indicates the gift. The question is: Does the Dative itself indicate the recipient or merely that the attached word is neither the giver nor the gift?

Diver makes the latter choice. In particular, he states that 'knowing that *mihi*, in the Dative, can be neither the Agent (the giver) nor the Patient (the gift), we deduce that it is the recipient'.

nominative and the other either dative or accusative, but typically accusative. The nominative here is the case of the agent, but this time the accusative (or the dative, whichever occurs) is the residue case. In a two-noun sentence, in other words, the accusative is not limited to the meaning of patient; it can express any number of other meanings as well. And, since it no longer contrasts with dative, it can be replaced by a dative. The choice between dative and accusative in two-noun sentences, since it is not semantically relevant, is subject to random kinds of free and conditioned variation.

Carrying the argument through, the noun found in a one-noun sentence can express any meaning relationship with the verb. The noun, though most frequently nominative, may be accusative or dative, but the choice is not based on meanings associated with these cases. When the noun is nominative its 'syntactic meaning' may be that of agent, patient, or anything else.

The inadequacy of Diver's treatment is clear. In the first place, it seems unlikely that, as used in his paper, the notions agent and patient are in any sense satisfactory semantic primitives. To agree that *imperium* in *senatus imperium mihi dedit* is the patient is nothing more than to agree to say the word 'patient' on seeing an accusative form in a three-noun sentence. For many of Diver's examples, his argument would have been every bit as convincing if he had said that an unvarying function is performed by the dative, but the role of the accusative depends on such matters as the lexical meaning of the verb. Furthermore, the 'couple of dozen verbs' which appear in two-noun sentences and which exhibit some kind of semantic correlation involving the supposedly nonsignificant choice of accusative or dative should probably not be set aside as unimportant exceptions.

Diver's proposal may be thought of as an attempt to identify the semantic contribution of cases seen as syntagmatically identified entities, while the positing of distinctive oppositions, in the manner of Hjelmslev and Jakobson, is an attempt to see the functioning of cases from the point of view of the concept of paradigmatic contrast. The latter view has been criticized by Kuryłowicz (1960, pp. 134, 141). The apparent contrast seen in Polish and Russian between accusative and genitive (partitive) direct object, as between 13 and 14

(13) *Daj nam chleb.* 'Give us the bread!'
(14) *Daj nam chleba.* 'Give us some bread!'

is not a difference in the syntactic function of the object nouns relative to the verb, but is rather a difference which falls into that area of syntax that deals with the effect of the choice of article, in languages having articles, on the semantic content of the associated noun. The fact that in Russian the difference is reflected as a difference in noun inflection does not alone determine its character as a part of the case system proper of the language.

The vertical contrast between locative and accusative nouns after locative/ directional prepositions, as in 15 and 16

(15) *On prygajet na stole.* 'He jumps (up and down) on the table.'
(16) *On prygajet na stol.* 'He jumps onto the table.'

is a difference that would be discussed in transformational grammar terms as involving a distinction between prepositional phrases which are inside and those which are outside the verb phrase constituent. That is, a locative prepositional phrase which occurs outside the constituent VP is one which indicates the place where the action described by the VP takes place. A locative prepositional phrase inside the VP is a complement to the verb. Inside a VP the difference between the locative and directional senses is entirely dependent on the associated verb; outside the VP the sense is always locative.

Kuryłowicz discussed 15 and 16 in essentially the same terms. To him the directional phrase *na stol* is 'more central' to the verb than the locative phrase *na stole.* An apparent contrast appears just in case the same verb may appear sometimes with and sometimes without a locative (or directional) complement. There is thus no genuine paradigmatic contrast in such pairs as 13–14 or 15–16.

Kuryłowicz's own approach to the study of case systems brings another order of grammatical fact into consideration: sentence relatedness. Cases, in his view, form a network of relationships mediated by such grammatical processes as the passive transformation. The distinction between nominative and accusative, for example, is a reflection in the case system of the more basic distinction between passive and active sentences. In his terms, *hostis occiditur* becomes the predicate *hostem occidit*, the primary change from *occiditur* to *occidit* bringing with it the concomitant change from *hostis* to *hostem.*

Nominalizations of sentences have the effect of relating both accusative and nominative to the genitive, for the former two are neutralized under conversion to genitive, as illustrated by the change from *plebs secedit* to *secessio plebis* (*genitivus subjectivus*) as opposed to the change from *hostem occidere* to *occisio hostis* (*genitivus objectivus*).

The relationship between nominative and accusative, then, is a reflex of diathesis; the relationship of these two to genitive is mediated through the process of constructing deverbal nouns. The remaining cases–dative, ablative, instrumental, and locative–enter the network of relationships in that, secondarily to their functions as adverbials, they each provide variants of the accusative with certain verbs. That is, there are verbs that 'govern' the ablative (for example, *utor*), rather than the accusative for their 'direct objects'.[15]

1.3. *Case Histories*

In addition to studies of case uses and interpretations of the cases in a given language as elements of a coherent system, the literature also contains many historical studies of cases; and these, too, are of various kinds. Some workers have sought to discover the original meanings of the cases of a language or family of languages, while others have sought to trace case morphemes back to other kinds of morphemes–either syntactic function words or some kind of derivational morphemes. Still others have seen in the history of one case system a case system of a different type–with or without assumptions concerning the 'essential primitivity' of the earlier type.

A very common assumption among linguistic historians has been that case affixes are traceable back to noncase notions. The form which eventually became the Indo-European case ending representing nominative singular masculine, that is, *-*s*, has been interpreted as the demonstrative **so* which had been converted into a suffix indicating a definite subject; and the **so* in turn is believed by some to have originated as a Proto-Indo-Hittite sentence connective (Lane, 1951). The same form has also been interpreted as a derivational morpheme indicating a specific individual directly involved in an activity, contrasting with a different derivational affix *-*m* indicating a nonactive object or the product of an action.[16] Scholars who can rest with the latter view are those who do not require of themselves the belief that 'synthetic' languages necessarily have antecedent 'analytic' stages.[17]

---

15. Kuryłowicz (1960, pp. 138–139, 144–147, 150). Also see Kuryłowicz (1964, pp. 179–181). Somewhat similar interpretations of the connections between case and diathesis are found in Heger (1966).

16. See, for example, the statement in Lehmann (1958, p. 190).

17. The impression is sometimes given that the identification of the etymon of a case affix brings with it an account of the intellectual evolution of the speakers of the language in question. If the interpretation of *-*m* and *-*s* as derivational morphemes is correct, it does not follow that one has discovered, in the transition from the earliest function of these elements to their later clear caselike uses, any kind of 'abstraction' process or tendency to pass from 'concrete' to 'relational' modes of thought. Our methods of reconstruction should certainly make it possible to detect basic (that is, deep-structure) linguistic evolution if it is there to discover, but the etymology of surface-structure

A second kind of speculation on historical changes within case systems traces case systems of one kind back to case systems of another kind. Of particular interest here is the suggestion that the Indo-European case systems point back to an original 'ergative' system. Case typologies will be discussed in slightly greater detail below, but briefly we can characterize an 'ergative' system as one which assigns one case (the ergative) to the subject of a transitive verb and another to both the subject of an intransitive verb and the object of a transitive verb. An 'accusative' system, on the other hand, is one which assigns one case to the subject of either transitive or intransitive verbs and another (the accusative) to the object of a transitive verb. A common feature of ergative systems is that the 'genitive' form is the same as the ergative (or, put differently, that the ergative case has a 'genitive' function).

The connection of Indo-European **-s* with animateness (the subject of a transitive verb is typically animate), the original identity of the nominative singular **-s* with the genitive ending, and the identity of the neuter ending **-m* with the masculine accusative form have led many investigators to the conclusion that our linguistic ancestors were speakers of an 'ergative' language.[18] It will be suggested below that, if such a change has taken place, it is a change which involves the notion 'subject'.

1.4. *Case in Current Generative Grammar*

A hitherto largely unquestioned assumption about case in the writings of generative grammarians has been made explicit by Lyons (1966, p. 218): '"case" (in the languages in which the category is to be found) is not present in "deep structure" at all, but is merely the inflexional "realization" of particu-

morphemes should not lead to assumptions about deep typological differences. What I mean is that the underlying case structures of Proto-Indo-European may have been just as precisely organized as those of any of the daughter languages, and that the changes that have occurred may have been entirely a matter of morphophonemic detail. From the preponderance of (derived) active nouns in subject position, one generation may have 'reinterpreted' the suffix as a marker of human subject and a later generation may have reinterpreted it as merely a marker for the subjectival use of a particular set of words–to state the possibilities in the most simple-minded way. The change, in short, may well have been entirely in the economies of bringing to the surface underlying structural features which themselves underwent no change whatever.

18. See particularly Uhlenbeck (1901), where the **-m* ending was identified as a subject marker and the **-s* as the agent marker in passive sentences (a common interpretation of 'ergative' systems), and Vaillant (1936). Lehmann (1958, p. 190) finds the arguments unconvincing, noting for example that evidence of an 'ergative' ending cannot be found in plural nouns or in *ā* stem feminines.

lar syntactic relationships'. The syntactic relationships in question may in fact be relationships that are defined only in the surface structure, as when the surface subject of a sentence (destined to assume, say, a 'nominative' form) has appeared as the result of the application of the passive transformation, or when the 'genitive' marker is introduced as an accompaniment to a nominalization transformation. One of Chomsky's few remarks on case occurs in a discussion of the peripheral nature of stylistic inversions; although case forms are assigned to English pronouns relatively late in the grammar, determined largely by surface-structure position, the stylistic inversion rules are later still. In this way it becomes possible to account for such forms as *him I like*; the shift of *him* to the front of the sentence must follow the assignment of case forms to the pronouns (see Chomsky, 1965, pp. 221 f.).

It seems to me that the discussion of case could be seen in a somewhat better perspective if the assignment of case forms were viewed as exactly analogous to the rules for assigning prepositions in English, or postpositions in Japanese.[19] There are languages which use case forms quite extensively, and the assumption that the case forms of nouns can be assigned in straightforward ways on the basis of simply defined syntactic relations seems to be based too much on the situation with English pronouns.

Prepositions in English—or the absence of a preposition before a noun phrase, which may be treated as corresponding to a zero or unmarked case affix—are selected on the basis of several types of structural features, and in ways that are exactly analogous to those which determine particular case forms in a language like Latin: identity as (surface) subject or object, occurrence after particular verbs, occurrence in construction with particular nouns, occurrence in particular constructions, and so on. The only difficulties in thinking of these two processes as analogous are that even the most elaborate case languages may also have combinations of, say, prepositions with case forms, and that some prepositions have independent semantic content. The first of these difficulties disappears if, after accepting the fact that the conditions for choosing prepositions are basically of the same type as those for choosing case forms, we merely agree that the determining conditions may simultaneously determine a preposition *and* a case form. The second difficulty means merely that a correct account will allow certain options in the choice of prepositions in some contexts, and that these choices

19. The suggestion is of course not novel. According to Hjelmslev, the first scholar to show a connection between prepositions and cases was A.-F. Bernhardi (1805); see Hjelmslev (1935, p. 24).

have semantic consequences. Analogous devices are provided by the 'true' case languages, too, for example by having alternative case choices in otherwise identical constructions, or by having semantically functioning prepositions or postpositions.

The syntactic relations that are involved in the selection of case forms (prepositions, affixes, and so forth) are, in practice, of two types, and we may call these 'pure' or 'configurational' relations, on the one hand, and 'labeled' or 'mediated' relations on the other hand.[20] 'Pure' relations are relations between grammatical constituents expressible in terms of (immediate) domination. Thus, the notion 'subject' can be identified as the relation between an NP and an immediately dominating S, while the notion 'direct object' can be equated with the relation that holds between an NP and an immediately dominating VP. Where the relation 'subject of' is understood to hold between elements of the deep structure, one speaks of the deep-structure subject; where it is understood to hold between elements of the (prestylistic) surface structure, one speaks of the surface-structure subject. This distinction appears to correspond to the traditional one between 'logical subject' and 'grammatical subject'.

By 'labeled' relation I mean the relation of an NP to a sentence, or to a VP, which is mediated by a pseudocategory label such as Manner, Extent, Location, Agent.

It is clear that if all transformations which create surface subjects have the effect of attaching an NP directly to an S, under conditions which guarantee that no other NP is also directly subjoined to the same S, and if it always turns out that only one NP is subjoined to a VP in the prestylistic surface structure, then these two 'pure' relations are exactly what determine the most typical

20. The distinction would be more accurately represented by the opposition 'relations' versus 'categories', because when a phrase-structure rule introduces a symbol like Manner or Extent–symbols which dominate manner adverbials and extent phrases–these symbols function, as far as the rest of the grammar is concerned, in exactly the same ways as such 'intentional' category symbols as S or NP. This fact has much more to do with the requirements of the phrase-structure model than with the 'categorial' character of the grammatical concepts involved. In an earlier paper I discussed the impossibility of capturing, in a base component of a grammar of the type presented in Chomsky (1965), both such information that *in a clumsy way* is a manner adverbial (and as such represents an instance of highly constrained lexical selection as well as a quite specific positional and co-occurrence potential which it shares with other manner adverbials) and that it is a prepositional phrase. See Fillmore (1966a).

The *intention* on the part of grammarians who have introduced such terms as Loc, Temp, Extent, and the like into their rules is to let these terms represent relations between the phrases they dominate and some other element of the sentence (that is, the VP as a whole); nobody, as far as I can tell, has actually wished these terms to be considered as representing distinct types of grammatical categories on the order of NP or preposition phrase.

occurrences of the case categories 'nominative' and 'accusative' in languages of a certain type. For remaining case forms, the determination is either on the basis of idiosyncratic properties of specific governing words, or on the basis of a 'labeled' relation, as when the choice of *by* is determined by reference to the dominating category Extent in the extent phrase of sentences like 17.

(17) He missed the target by two miles.

In my earlier paper (Fillmore, 1966a) I pointed out that no semantically constant value is associated with the notion 'subject of' (unless it is possible to make sense of the expression 'the thing being talked about', and, if that can be done, to determine whether such a concept has any connection with the relation 'subject'), and that no semantically relevant relations reside in the surface subject relation which are not somewhere also expressible by 'labeled' relations. The conclusion I have drawn from this is that all semantically relevant syntactic relations between NPs and the structures which contain them must be of the 'labeled' type. The consequences of this decision include (*a*) the elimination of the category VP, and (*b*) the addition to some grammars of a rule, or system of rules, for creating 'subjects'. The relation 'subject', in other words, is now seen as exclusively a surface-structure phenomenon.

2. Some Preliminary Conclusions

I have suggested that there are reasons for questioning the deep-structure validity of the traditional division between subject and predicate, a division which is assumed by some to underlie the basic form of all sentences in all languages. The position I take seems to be in agreement with that of Tesnière (1959, pp. 103–105) who holds that the subject/predicate division is an importation into linguistic theory from formal logic of a concept which is not supported by the facts of language and, furthermore, that the division actually obscures the many structural parallels between 'subjects' and 'objects'. The kinds of observations that some scholars have made about surface differences between 'predicative' and 'determinative syntagms'[21] may be accepted without in any way believing that the subject/predicate division plays a part in the *deep-structure* syntactic relations among the constituents of sentences.

21. See, for example, Bazell (1949, esp. p. 8), where the difference is expressed in such terms as 'degrees of cohesion', 'liaison features' found within the predicate but not between subject and predicate.

Once we have interpreted 'subject' as an aspect of the surface structure, claims about 'subjectless' sentences in languages which have superficial subjects in some sentences, or reports about languages which appear to lack entirely entities corresponding to the 'subjects' of our grammatical tradition, no longer need to be regarded as particularly disturbing. Unfortunately, there are both good and bad reasons for asserting that particular languages or particular sentences are 'subjectless', and it may be necessary to make clear just what I am claiming. A distinction must be drawn between *not having* a constituent which could properly be called 'subject', on the one hand, and *losing* such a constituent by anaphoric deletion, on the other hand.[22] Robins (1961), in his review of Tesnière (1959), accuses Tesnière of failing to isolate the subject from the rest of the sentence. To Robins, Tesnière's decision to allow the subject to be treated as merely a complement to the verb must be related to the fact that the subject is omissible in such languages as Latin. If it is true that the *omissibility* of subjects is what convinced Tesnière that they are subordinated to verbs, and if the nonomissibility in any language of the subject constituent would have persuaded him that there *is* a special status for 'subject' vis-à-vis 'predicate' in the underlying structure of sentences in all languages, then that, it seems to me, is a bad reason for coming up with what might be a correct analysis.

It seems best to have a place in linguistic theory for the operation of anaphoric processes, processes which have the effect of shortening, simplifying, de-stressing sentences which are partly identical to their neighbors (or which are partly 'understood'). It happens that English anaphoric processes make use of pronominalization, stress reduction, and also deletion, under condi-

22. The tagmemicists in particular, because of their notation for optional constituents, have had to come to grips with this distinction. A 'tagmemic formula' may be thought of as an attempt to present in a single statement a quasi-generative rule for producing a set of related sentences *and* the surface structure (short of free variation in word order) of these sentences. If the formulas for transitive and intransitive clauses are expressed as i and ii respectively:

(i) ± Subj ± Pred ± Obj; ± Loc ± Time
(ii) ± Sub; ± Pred ± Loc ± Time

it is clear (*a*) that any clause containing just a Pred can satisfy either of these formulas, and (*b*) that the potential appearance of such constituents as Loc and Time is less relevant to the description of these clauses than is that of the constituent Obj. Pike draws a distinction, which cross-cuts the optional/obligatory distinction, between 'diagnostic' and 'nondiagnostic' elements of clauses; see, for example, Pike (1966, esp. Chapter I, Clauses). Grimes, on the other hand, seems to suggest introducing the 'diagnostic' constituents obligatorily, allowing for their deletion under certain contextual or anaphoric conditions. See Grimes (1964, esp. p. 16 f.).

tions where other languages might get along exclusively with deletion.[23] Under some conditions, in languages of the latter type, the deleted element happens to be the 'subject'. The nonoccurrence of subject nouns in some utterances in some languages is *not* by itself, in other words, a good argument against the universality of the subject/predicate division. There are better ones. Some of these have already been suggested, others are to appear shortly.

By distinguishing between surface- and deep-structure case relationships, by interpreting the 'subject' and 'object' as aspects of the surface structure, and by viewing the specific phonetic shapes of nouns in actual utterances as determinable by many factors that are vastly variable in space and time, we have eliminated reasons for being surprised at the noncomparability of (surface) case systems. We find it partly possible to agree with Bennett (1914, p. 3) when, after surveying a few representative nineteenth century case theories, he stated that they erred in sharing the 'doubtful assumption ... that all the cases must belong to a single scheme, as though parts of some consistent institution'. We need not follow him, however, in concluding that the only valid type of research into the cases is an inquiry into the earliest value of each case.

Greenberg (1966, p. 98; see also p. 80) has remarked that cases themselves cannot be compared across languages–two case systems may have different numbers of cases the names of the cases may conceal functional differences–but that *case uses* may be expected to be comparable. He predicts, for instance, that the uses of cases will be 'substantially similar in frequency but differently combined in different languages'. Greenberg's recommendations on the crosslinguistic study of case uses were presented in connection with the 'true' case languages, but it seems clear that if a 'dative of personal agent' in one language can be identified with an 'ablative of personal agent' in another language, then the 'personal agent' relationship between a noun and a verb ought also to be recognizable in the so-called caseless languages on exactly the same grounds. If, furthermore, it turns out that other grammatical facts can be associated with sentences containing the personal agent relationship, it would appear that the concepts underlying the study of case uses may have a greater linguistic significance than those involved in the description of surface case systems. These additional facts might include the identification of a limited set of nouns and a limited set of verbs capable of entering into this relationship, and whatever additional generalizations prove to be statable in terms of this classification. Higher level dependencies may be discovered,

23. For an extremely informative description of these processes in English, see Gleitman (1965) and Harris (1957, esp. section 16).

such as the limitation of benefactive phrases to sentences containing a personal agent relationship in their deep structure.

The question should now be asked, of course, whether we are justified in using the term *case* for the kind of remote syntactic-semantic relations that are at issue. There is among many scholars a strong feeling that the term should be used only where clear case morphemes are discoverable in the inflection of nouns. To Jespersen (1924, p. 186), it is wrong to speak of 'analytic' cases, even when there is no 'local' meaning in the preposition phrases, because cases are one thing and preposition-plus-object constructions are another. Jespersen's position is colored a little by his belief that the caselessness of English represents a state of progress for which we ought to be grateful.[24]

Cassidy (1937, p. 244), in his appeal to rescue the word *case* from abuse, wrote: '"Case" will be properly used and will continue to have some meaning only if the association with inflection be fully recognized, and if stretching of the term to include other sorts of "formal" distinction be abandoned.' In a similar vein, Lehmann (1958, p. 185) chides Hirt for suggesting that an awareness of cases had to precede the development of case endings–that there was, in other words, 'among the speakers of Pre-Indo-European and Proto-Indo-European a disposition for cases'. Lehmann continues: 'We can account for Hirt's statement by the assumption that to him a case was a notional category, whether or not it was exemplified in a form. To us a particular case is non-existent unless it is represented by forms which contrast in a system with others.' The claim that syntactic relations of various types must exist before case endings could be introduced to give them expression would surely have gone unchallenged; what was offensive, apparently, was the use of the word *case*.

It seems to me that if there are recognizable intrasentence relationships of the types discussed in studies of case systems (whether they are reflected in case affixes or not), that if these same relationships can be shown to be comparable across languages, and that if there is some predictive or explanatory use to which assumptions concerning the universality of these relations can be put, then surely there can be no meaningful objection to using the word

---

24. Jespersen (1924, p. 179):

> However far back we go, we nowhere find a case with only one well-defined function: in every language every case served different purposes, and the boundaries between these are far from being clear-cut. This, in connection with irregularities and inconsistencies in the formal elements characterizing the cases, serves to explain the numerous coalescences we witness in linguistic history ("syncretism") and the chaotic rules which even thus are to a great extent historically inexplicable. *If the English language has gone farther than the others in simplifying these rules, we should be devoutly grateful and not go out of our way to force it back into the disorder and complexity of centuries ago.* [Italics added.]

case, in a clearly understood deep-structure sense, to identify these relationships. The dispute on the term case loses its force in a linguistics which accepts the centrality of syntax.[25]

We may agree, then, for our present purposes, with Hjelmslev (1935), who suggests that the study of cases can be pursued most fruitfully if we abandon the assumption that an essential characteristic of the grammatical category of case is *expression in the form of affixes on substantives.* I shall adopt the usage first proposed, as far as I can tell, by Blake (1930) of using the term *case* to identify the underlying syntactic-semantic relationship, and the term *case form* to mean the expression of a case relationship in a particular language–through affixation, suppletion, use of clitic particles, or constraints on word order.

3. Case Grammar

The substantive modification to the theory of transformational grammar which I wish to propose amounts to a reintroduction of the 'conceptual framework' interpretation of case systems, but this time with a clear understanding of the difference between deep and surface structure. The sentence in its basic structure consists of a verb and one or more noun phrases, each associated with the verb in a particular case relationship. The 'explanatory' use of this framework resides in the necessary claim that, although there can be compound instances of a single case (through noun phrase conjunction), each case relationship occurs only once in a simple sentence.[26]

It is important to realize that the explanatory value of a universal system of deep-structure cases is of a syntactic and not (merely) a morphological nature. The various permitted arrays of distinct cases occurring in simple

25. The universality of case as a grammatical category is affirmed in Hjelmslev (1935, p. 1). In a recent study from a Jakobsonian point of view, Velten (1962) reveals enough of the historical continuity of 'synthetic' and 'analytic' cases to suggest that the linguist has no right to assign cases and prepositions to different 'chapters' of the study of grammar. The deep-structure notion of cases may be thought of as involving an extension of the synchronic concept of 'syncretism'. The usual synchronic sense of case syncretism assumes the form of a decision to posit a case contrast that may not be expressed overtly in most contexts as long as it appears overtly in 'one part of the system'. (See Newmark 1962, p. 313.) Deep-structure cases may simply be *nowhere* overtly reflected as affixes or function words. The notion we are after probably corresponds to Meinhof's *Kasusbeziehungen.* (See Meinhof 1938, p. 71.) The Meinhof reference, which I have not seen, was quoted in Frei (1954, fn. p. 31).

26. It follows that whenever more than one case form appears in the surface structure of the same sentence (on different noun phrases), either more than one deep-structure case is involved or the sentence is complex. If, for example, German *lehren* is described as a verb which 'takes two accusatives', we have reason to believe that in the deep structure, the two object nouns are distinct as to case. Often enough the language will provide evidence for the distinction, as in the occurrence of such passive sentences as *das wurde mir gelehrt.*

sentences express a notion of 'sentence type' that may be expected to have universal validity, independently of such superficial differences as subject selection. The arrays of cases defining the sentence types of a language have the effect of imposing a classification of the verbs in the language (according to the sentence type into which they may be inserted), and it is very likely that many aspects of this classification will be universally valid.

Case elements which are optionally associated with specific verbs, together with the rules for forming subjects, will serve to explain various co-occurrence restrictions. For example, in 18 the subject is in an Agent relation to the verb; in 19 the subject is an Instrument; and in 20 both Agent and Instrument appear in the same sentence, but in this case it is the Agent which appears as the subject, not the Instrument.

(18) John broke the window.
(19) A hammer broke the window.
(20) John broke the window with a hammer.

That the subjects of 18 and 19 are grammatically different explains the fact that the combined meaning of the two sentences is not produced by conjoining their subjects. Thus 21 is unacceptable.

(21) *John and a hammer broke the window.

Only noun phrases representing the same case may be conjoined. Similarly, the fact that only one representative of a given case relationship may appear in the same simple sentence, together with the generalizations on subject selection and the redundancies which hold between cases and lexical features (for example, between Agent and animateness), explains the unacceptability of sentence 22.

(22) *A hammer broke the glass with a chisel.

It is unacceptable, in particular, on the interpretation that both *hammer* and *chisel* are understood instrumentally. It cannot represent a sentence containing an Agent and an Instrument, since the noun *hammer* is inanimate.[27]

---

27. The author is aware that in sentence 18 one might be talking about what John's body did as it was tossed through the window and that in sentence 19 one might be speaking metaphorically, personifying *hammer.* Under either interpretation sentence 21 turns out to be acceptable, and under the personification interpretation, sentence 22 becomes acceptable. What is important to realize is that these interpretations, too, are explainable by reference to exactly the same assumptions appealed to in explaining their 'face value' interpretations.

The dependency that can be accounted for by making these assumptions is that the subject of an active transitive sentence must be interpretable as a personal agent just in case the sentence contains a *with* phrase of instrumental import. Apparent exceptions to this generalization can be seen to have different underlying structures. sentence 23 looks like an exception, but by attending to the effect of the word *its*, the essential difference between 23 and sentences 22 and 24 becomes apparent.

(23) The car broke the window with its fender.
(24) *The car broke the window with a fender.

Sentence 24 violates the conditions that have been discussed, but sentence 23 is a paraphrase of sentence 25 and may be interpreted as having the same structure as 25.

(25) The car's fender broke the window.

What is suggested here is that sentences 23 and 25 are agentless sentences containing *a possessed noun* as the Instrument (*the car's fender*). The rules for choosing a subject allow an option in this case: either the entire instrument phrase may appear as the subject (as in 25), or the 'possessor' alone may be made the subject, the remainder of the instrument phrase appearing with the preposition *with* (as in 23). The second option requires that a 'trace' be left behind in the instrument phrase, in the form of the appropriate possessive pronoun. A similar explanation is suggested for such sentences as 26 and 27, which are also interpretable as deep structurally identical.

(26) Your speech impressed us with its brevity.
(27) The brevity of your speech impressed us.

The superficial nature of the notion 'subject of a sentence' is made apparent by these examples in a particularly persuasive way, because in the possessor-as-subject cases, the 'subject' is not even a major constituent of the sentence; it is taken from the modifier of one of the major constituents.

In the basic structure of sentences, then, we find what might be called the 'proposition', a tenseless set of relationships involving verbs and nouns (and embedded sentences, if there are any), separated from what might be called the 'modality' constituent. This latter will include such modalities on the sentence-as-a-whole as negation, tense, mood, and aspect.[28] The exact nature of the modality constituent may be ignored for our purposes. It is likely, however, that certain 'cases' will be directly related to the modality constituent as others are related to the proposition itself, as for example certain temporal adverbs.[29]

The first base rule, then, is 28, abbreviated to 28'.

(28) Sentence → Modality + Proposition
(28') S → M + P[30]

The P constituent is 'expanded' as a verb and one or more case categories. A later rule will automatically provide for each of the cases the categorial realization as NP (except for one which may be an embedded S). In effect the case relations are represented by means of dominating category symbols.

The expansion of P may be thought of as a list of formulas of the form seen in 29, where at least one case category must be chosen and where no case category appears more than once.

(29) $P \rightarrow V + C_1 + \ldots + C_n$

Whether these formulas can be collapsed according to the familiar abbreviatory conventions is not at present clear. For our purposes we may simply think of P as representable by any of a set of formulas including V+ A, V+ O+ A, V+ D, V + O + I +A, and so forth. (The letter symbols are interpreted below.)

The case notions comprise a set of universal, presumably innate, concepts which identify certain types of judgments human beings are capable of making about the events that are going on around them, judgments about such

28. There are probably good reasons for regarding negation, tense, and mood as associated directly with the sentence as a whole, and the perfect and progressive 'aspects' as features on the V. See for a statement of this position Lyons (1966, pp. 218, 223).

29. In my earlier paper I suggested that sentence adverbials in general are assigned to the modality constituent. I now believe that many sentence adverbs are introduced from superordinate sentences (by transformations of a type we may wish to call 'infrajections'). This possibility has long been clear for unmistakable sentence adverbs like *unfortunately*, but there are also quite convincing reasons for extending the infrajection interpretation to adverbs like *willingly*, *easily*, and *carefully*.

30. The arrow notation is used throughout, but this should not be interpreted as meaning that the proposal for a case grammar requires an assumption of a left-to-right orientation of the constituent symbols of the rewriting rules.

matters as who did it, who it happened to, and what got changed. The cases that appear to be needed include:

Agentive (A), the case of the typically animate perceived instigator of the action identified by the verb.[31]

Instrumental (I), the case of the inanimate force or object causally involved in the action or state identified by the verb.[32]

Dative (D), the case of the animate being affected by the state or action identified by the verb.

Factitive (F), the case of the object or being resulting from the action or state identified by the verb, or understood as a part of the meaning of the verb.

Locative (L), the case which identifies the location or spatial orientation of the state or action identified by the verb.

Objective (O), the semantically most neutral case, the case of anything representable by a noun whose role in the action or state identified by the verb is identified by the semantic interpretation of the verb itself; conceivably the concept should be limited to things which are affected by the action or state identified by the verb.[33] The term is not to be confused with the notion of direct object, nor with the name of the surface case synonymous with accusative.

Additional cases will surely be needed. Suggestions for adding to this list will appear in various places below.

It is important to notice that none of these cases can be interpreted as

---

31. The escape qualification 'typically' expresses my awareness that contexts which I will say require agents are sometimes occupied by 'inanimate' nouns like *robot* or 'human institution' nouns like *nation*. Since I know of no way of dealing with these matters at the moment, I shall just assume for all agents that they are 'animate'.

32. Paul Postal has reminded me of the existence or sentences like

(i) I rapped him on the head with a snake.

The requirement that instrumental NP's are inanimate is the requirement to interpret i as having in its underlying structure something equivalent to *with the body of a snake*. The fact that there are languages which would require mention of a stem meaning 'body' in this context may be considered as support for this position, and so may the unacceptability, pointed out by Lakoff, of sentences like ii:

(ii) *John broke the window with himself. (See Lakoff 1967.)

33. In Fillmore (1966a) the neutral case was unwisely and misleadingly labeled 'ergative'.

matched by the surface structure relations, subject and object, in any particular language. Thus, *John* is A in 30 as much as in 31; *the key* is I in 32 as well as in 33 or 34; *John* is D in 35 as well as in 36 and 37; and *Chicago* is L in both 38 and 39.

(30) John opened the door.
(31) The door was opened by John.
(32) The key opened the door.
(33) John opened the door with the key.
(34) John used the key to open the door.
(35) John believed that he would win.
(36) We persuaded John that he would win.
(37) It was apparent to John that he would win.
(38) Chicago is windy.
(39) It is windy in Chicago.

The list of cases includes L, but nothing corresponding to what might be called directional. There is a certain amount of evidence, as was mentioned above, that locational and directional elements do not contrast but are superficial differences determined either by the constituent structure or by the character of the associated verb. An example provided by Hall (40) suggests, by the occurrence of the pro replacement word *there*, that *to the store* and *at the store* are variants of the same entity, determined by the movement or non-movement character of the associated verb.[34]

---

34. Hall (1965).

The putative contrast between locational and directional expressions as well as the distinction between 'optional' and 'obligatory' locative expressions, as exemplified in Hall's examples i and ii, seem to point to the difference between elements which are 'inside the VP' and elements which are 'outside the VP'.

(i) John keeps his car in the garage.
(ii) John washes his car in the garage.

In our terms this would be equivalent either to determining whether there is a difference between an L as a constituent of P and an L as a constituent of M, or whether there can be two L elements within P, distinguished in terms of degree of selectivity of verbs. The highly restricting L selects verbs like *keep, put,* and *leave,* but not *polish, wash,* and *build*; the weakly restricting L selects verbs like *polish, wash,* and *build,* but not *believe, know,* or *want.*

However this distinction is interpreted, the second or 'outer' L is in some respects similar in its'selectional' properties to what might be called the *benefactive* case B. B, too, is involved in the selection of verbs in the sense that some verbs do not accept B modification (*'He is tall for you'); but the restriction here may have more to do with *dependency relations between cases* than with

(40) She took him to the store and left him there.

I have stated that A and D are 'animate' participants in the activity of the associated verbs, and I have also suggested that verbs are selected according to the case environments the sentence provides–what I shall refer to as the 'case frame'. There are, then, the two problems of lexical selection, that of the nouns and that of the verbs. Those features of nouns required by a particular case are to be specified by obligatory rules of the type such as the following, which specifies that any N in an A or D phrase must contain the feature [+animate]. (Recall the qualification of footnote 30.)

$$N \rightarrow [+\text{animate}]/^{A,\,D}\,[X\_\_\_\_Y]$$

To take care quite generally of lexical features associated with specific cases, we may appeal to a rule which associates with each noun a label identifying the case relation it holds with the rest of the sentence. Such a rule might associate with every noun under L the feature [+locative], for example, Since abstract nouns such as *idea* cannot serve as heads of L expressions, they will be marked [–locative].[35]

The insertion of verbs, on the other hand, depends on the particular array of cases, the 'case frame', provided by the sentence.[36] The verb *run*, for exam-

---

dependencies directly connected with the verb. It appears, in fact, that those verbs which allow 'outer L' and B modification are precisely those which take agents. I have no ideas on how these dependencies can be stated, but it would appear that the second L and the B can appear only in sentences containing As.

Thus the *regime direct* versus *regime indirect* interpretation of the difference between iii and iv

(iii) *Il demeure à Paris.*
(iv) *Il travaille à Paris.*

may have simply to do with the fact that the subject of iv is actually an A. Both the specific verb and the occurrence of an 'outer L' are determined by the presence of an A. See, in this connection, Bazell's discussion (1949, p. 10) of Gougenheim's review of de Boer's French syntax.

35. By allowing highly restricting lexical features to be associated with given case units we have returned to that extension of 'cases' to 'adverb forms' proposed by Bopp, Wüllner, and Hartung. Some adverbs, on this view, are really nouns capable of 'taking' only one case form. Since deep structure cases are in fact all 'defective' to some extent, with respect to the nouns which they accept, such a concept as *inflectional scope* no longer provides a clear demarcation between 'case forms proper' and 'adverbs'. See the discussion of this question in Hjelmslev (1935, p. 40).

36. I am adhering, in this discussion, to the Postal-Lakoff doctrine, which I find thoroughly convincing, that adjectives constitute a subset of verbs.

ple, may be inserted into the frame [____A], the verb *sad* into the frame [____D], verbs like *remove* and *open* into [____O + A], verbs like *murder* and *terrorize* (that is, verbs requiring 'animate subject' and 'animate object') into [____D + A], verbs like *give* into [____O + D + A], and so on.

In lexical entries for verbs, abbreviated statements called 'frame features' will indicate the set of case frames into which the given verbs may be inserted. These frame features have the effect of imposing a classification of the verbs in the language. Such a classification is complex no1t only because of the variety of case environments possible within P but also because many verbs are capable of occurring in more than one distinct case environment. This last fact can be represented most directly by allowing facultative representation of cases in the frame-feature expressions.

The word *open*, to take a familiar example, can occur in [____O], as in 41; in [____O + A], as in 42; in [____O + I], as in 43; and in [____O + I + A], as in 44.

(41) The door opened.
(42) John opened the door,
(43) The wind opened the door.
(44) John opened the door with a chisel.

The simplest representation of this set of possibilities makes use of parentheses to indicate the 'optional' elements. The frame feature for *open* may thus be represented as 45.

(45) +[____O (I) (A)][37]

Other verbs having this same feature are *turn*, *move*, *rotate*, and *bend*.

For a verb like *kill* it is necessary to indicate, expressing it in familiar terms, that it takes an animate object and either an animate or an inanimate subject, and that if there is an animate subject, an instrument phrase may also co-occur. The frame feature for *kill*, in other words, will have to specify that either an Instrument or an Agent must be specified, and both may be specified. If the linked parentheses notation can be introduced to indicate that at

37. Case frames are represented in square brackets, with 'underline' indicating the position of the element with respect to which the expression is an environmental frame.

A frame feature is represented in square brackets with '+' or '–' in front, indicating that the set of case frames represented by the expression within the brackets is that which will (if the feature is marked '+') or which will not (if the feature is marked '–') accept the lexical item with which the feature is associated.

least one of the linked elements must be chosen, the frame feature for *kill* can be given as 46.

(46) + [_____ D (I)(A)]

The verb *murder*, on the other hand, is one which requires an Agent. Its frame feature differs from that of 45 and 46 because the element A is obligatorily present. It is given as 47.

(47) +[_____D (I) A]

The environmental subclassification of verbs is sensitive to more than the mere array of cases in P. Since one of the cases may be represented by S (an embedded sentence), verbs are also subclassified in terms of whether the O element is a sentence. By convention we shall interpret the symbol O in frame features as indicating NPs, and the symbol S as indicating an O to which an S has been embedded.

The frame feature +[_____S] characterizes such verbs as *true*, *interesting*, and so forth; the feature +[_____S + D] is common to such verbs as *want* and *expect*; verbs like *say*, *predict*, and *cause* appear in the frame [_____S + A]; and verbs like *force* and *persuade* are insertable into the frame [_____S + D + A].[38]

Verbs are distinguished from each other not only by specification of the case frames into which they can be inserted, but also by their transformational properties. The most important variables here include (*a*) the choice of a particular NP to become the surface subject, or the surface object, wherever these choices are not determined by a general rule; (*b*) the choice of prepositions to go with each case element, where these are determined by idiosyncratic properties of the verb rather than by a general rule; and (*c*) other special transformational features, such as, for verbs taking S complements, the choice of specific complementizers (*that*, *-ing*, *for to*, and so forth) and the later transformational treatment of these elements.

The use of parentheses in expressing the frame features, together with the transformational introduction of subjects, makes it possible to reduce the number of semantic descriptions in the lexicon. The semantic interpretation of a P will introduce all information provided by specific case relationships

---

38. It should be pointed out that descriptions of embedded sentences as *it* + S realizations of the category NP in 'subject/object' grammars must somehow guarantee that this particular expansion of NP is limited to the subjects of intransitive sentences and the objects (direct or oblique) of transitive sentences. All such restrictions are rendered unnecessary by the decision to limit complement S to the case element O.

represented in the P, allowing such information to be omitted from the semantic descriptions of verbs. In the case of verbs having the feature 45, as we have seen, certain related transitive and intransitive verbs need not be given separate semantic description. This point may be further demonstrated with the English verb *cook*. The frame feature of *cook* is presumably something like 48

(48) + [____O (A)]

and an idiosyncratic transformational feature of the verb is that just in case the A is present and the O is some NP representing a typical NP for the verb (that is, something like *food* or a *meal*), the O element may be deleted. The semantic description of the verb will do no more than identify a particular activity having a result of a particular kind on the object identified by the O element. The same semantic entry, in other words, will account for the use of cook in all of the sentences 49–51.

(49) Mother is cooking the potatoes.
(50) The potatoes are cooking.
(51) Mother is cooking.

Instead of saying that the verb has three different meanings, we can be satisfied to say that there is a certain variety in the case frames which accept it, and that it is one of the 'deletable object' verbs. The fact that A is obligatorily animate and that O is unspecified for animateness accounts for the fact that if we can read sentence 50 as ambiguous, it is because we can accept certain violations of grammatical requirements in 'personifications' of the type we have learned in nursery school, whereas if we accept 51 as in fact ambiguous, it is because we are acquainted with the range of activities found in human societies.

The example with *cook* shows that the lexicon need not contain as many semantic entry *tokens* under the present proposal as it would in a subject/object grammar.[39] It will now be shown that this same flexibility makes it

39. It may appear that facultative representation of cases in frame features has the advantages it does in English because there are so many verbs which can be used transitively or intransitively in the same form. It is a language-particular coincidence that English uses the same form in these words. The identification of transitive and intransitive *open*, or transitive and intransitive *cook*, is justified because the semantic characterization of the verb is the same in all of the uses discussed. (We must distinguish between the semantic characterization of a verb and the semantic interpre-

possible to reduce the number of semantic entry *types*, for now it is feasible to show that some syntactically different words are in fact semantically identical (with respect to that aspect of their meanings which is independent of the contribution of the associated cases). This may be true for verbs like *like* and *please*, to give the example that comes most quickly to mind. These words may be described as being synonymous. Each has the frame feature +[__ O + D]; they differ only in their subject selection features. The verb *like*, in fact, has in its history the subject selection feature possessed by *please*.

The verb *show*, to give another kind of example, might well have the same semantic representation as *see*, differing from it only in that the frame feature for *show* contains an A where that for *see* does not. The verbs *kill* and *die* appear to be related in a similar way.

(52) *see* (+[____O + D]) versus *show* (+[____O + D + A])
(53) *die* (+[____D]) versus *kill* (+[____D (I)A]

We have seen, then, instances of synonymy where there are identical frame features but different subject selection features, and instances of synonymy where there are frame feature differences depending on whether a particular case category was present or absent. We may now turn to examples of synonymy where the difference is in the choice of one case or another.

It will be recalled that both A and D are animate. The semantic descriptions of certain verbs may refer to the animateness of the associated noun, independently of whether the 'source' of the animateness is A or D. That is, the semantic representation of certain verbs may specify a relationship or a process associated with the necessarily animate participant in the state or activity identified by the verb. The relation of *hear* and *listen* to the necessarily animate NP is the same in both cases; the difference in the semantic interpretation of the Ps containing them is determined by the semantic contribution

---

tation of sentences containing the verb. In the latter case, all of the co-constituents and the semantic role they play as determined by their cases are taken into account.) Wherever that condition can be satisfied, facultative representation is called for. It will turn out that for some languages the occurrence or nonoccurrence of one of the 'optional' cases will have an effect on the verb. If, for verbs of the type [____O (A)], the appearance of the A determines a variant of the verb different from that when A is missing (distinguishing the 'transitive' from the 'intransitive' use of the 'same' verb), or if the absence of the A requires some additive element (for example, a 'reflexive' morpheme) not needed when the A is expressed, these facts can be provided transformationally. See Hashimoto (1966). (By extending the range of acceptable surface variants of verbs under these conditions to suppletion, it may even be possible to interpret the contrasts exemplified in 52–54 below as surface lexical variation.)

of the associated cases and by the fact that the frames that contain *hear* are [____O + D] and those that contain *listen* are [____O + A]. The fact that in the case of *listen* the relationship is understood as involving the active participation of the person identified as A is due to the presence of A, not to a special meaning of *listen*. The same distinction can be seen between *see* and *know*, on the one hand, and *look* and *learn*, on the other.

(54) *see, know* (+[____O + D]) versus *look, learn* (+[____O + A])

This latest point leads one to those properties of English verbs with which Lakoff (1966) associates the terms 'stative' and 'nonstative'. The question we need to ask is whether Lakoff's features are primitives in the lexical entries for verbs, or whether they permit reduction to concepts of the type I have been outlining. Lakoff has noticed that the 'true imperative', the progressive aspect, the occurrence of benefactive (B) phrases, and *do so* substitution occur only with 'nonstative' verbs. His discussion suggests that one must assign 'stative' and 'nonstative' as features on verbs and then guarantee that B phrases are permitted only with 'nonstatives' (put the other way around, one must guarantee that the presence of a B expression allows only for the selection of 'nonstatives'), that the imperative transformation can be applied only if the verb is 'nonstative', and so on. The treatment that I prefer is implicit in what I have already presented. The transformation which accounts for the 'true imperatives' can apply only to sentences containing As, and the occurrence of B expressions (and 'outer Ls') is dependent on the presence of an A. The progressive aspect can only be chosen in association with particular case frames, for example, those containing As. No special features indicating stativity need be added to verbs because, if this suggestion is correct, only those verbs which occur in Ps containing As will show up in these sentences anyway.[40]

3.1. *Surface Phenomena*

To recapitulate, our discussion so far has suggested that the deep structure of (the propositional component of) every simple sentence is an array consisting of a V plus a number of NPs holding special labeled relations (cases) to the sentence. These relations, which are provided for categorially, include such concepts as Agentive, Instrumental, Objective, Factitive, Locative, Benefactive, and perhaps several others. Complex sentences involve recursion

40. The *do-so* evidence is not so easy to interpret in this way. Still, the connection between 'nonstative' verbs and verbs that can 'take' A is too compelling to be simply wrong.

through the category Sentence under the case category Objective. Verbs are subclassified according to the case environments which accept them, and the semantic characterizations of verbs relate them either to specific case elements in the environment or to elements containing features (such as animateness) introduced as obligatory accompaniments of particular cases.

This section will deal with some of the ways in which deep structures of the type proposed in this essay are converted into surface representations of sentences. The various mechanisms involve selection of overt case forms (by suppletion, affixation, addition of prepositions or postpositions), 'registration' of particular elements in the verb, subjectivalization, objectivalization, sequential ordering, and nominalizations.

A surface case system may be related to the set of underlying cases in a variety of ways. Two deep cases may be represented in the same way in the surface structure, as when D and O direct objects are both represented with the 'accusative' case in many languages (where the determining factor may be occurrence immediately after the verb at some stage of derivation). A and D may be represented by the same overt form, where the determining factor may be case-linked animateness. Or the superficial form of a case element may be determined by an idiosyncratic property of some governing word.

The rules for English prepositions may look something like this: the A preposition is *by*; the I preposition is *by* if there is no A, otherwise it is *with*; the O and F prepositions are typically *zero*; the B preposition is *for*; the D preposition is typically *to*; the L and T (for time) prepositions are either semantically nonempty (in which case they are introduced as optional choices from the lexicon), or they are selected by the particular associated noun [*on the street*, *at the corner* (=intersection of two streets), *in the corner* (of a room); *on Monday*, *at noon*, *in the afternoon*]. Specific verbs may have associated with them certain requirements for preposition choice that are exceptions to the above generalization.[41]

The position of prepositions can be guaranteed either by having the case categories rewritten as *Prep* + NP, or by having *Prep* be one of the obligatory constituents of NP. I shall make the former choice, although the grounds for deciding one way or the other are not particularly clear. The 'universal' char-

---

41. The verb *blame*, for example, chooses ('governs') *for* for O and *on* for D. The O preposition is *at* for *look* meaning 'examine', *for* for *look* meaning 'seek', *to* for *listen*, and so forth. Changes in the original preposition assignment may be brought about by transformations: the rules which provide surface subjects and direct objects delete prepositions (replace them by zero), and the rules which form deverbal (= desentential) nominals convert some of the original case forms into 'genitive', either by replacing the assigned preposition with *of*, or, in some cases, by removing the original preposition and affixing the 'genitive' suffix.

acter of the base rules is kept intact by the assumption that prepositions, postpositions, and case affixes–semantically relevant or not–are all in fact realizations of the same underlying element, say *K* (for *Kasus*). We may regard all of the case categories as therefore rewritten as *K* + *NP*.

Every English sentence has a surface subject, if only formally so. For most combinations of cases there is a 'preferred' or 'unmarked' subject choice; for some there is no actual choice–the subject is uniquely determined. In general the 'unmarked' subject choice seems to follow the following rule:

(55) If there is an A, it becomes the subject; otherwise, if there is an I, it becomes the subject; otherwise, the subject is the O.

Suppose, for example, that the base representation of a particular sentence is item 56:

(56)

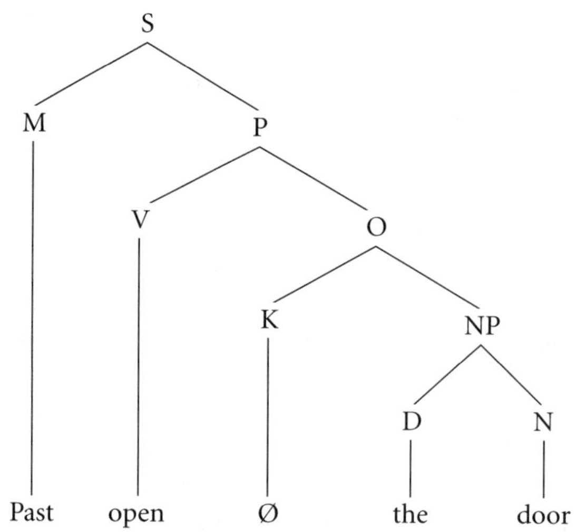

Since the sentence contains only one case category, it is obligatorily moved to the front (and hence directly subjoined to the category S) where it will later undergo subject-preposition deletion. There is a stage, in other words, where the form of the sentence in question is that represented in 57.

(57)

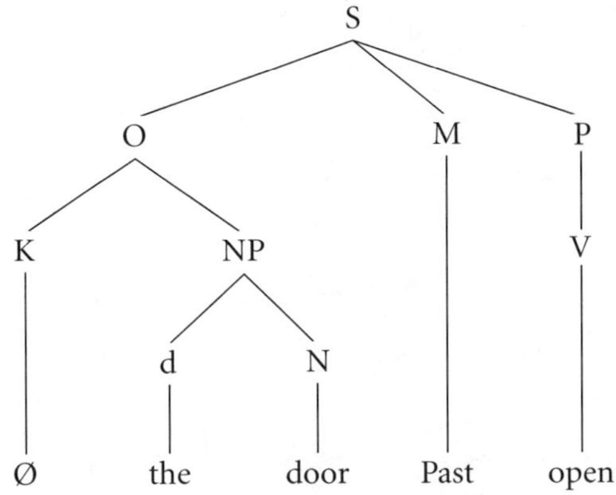

The subject-preposition deletion rule removes the preposition and deletes the case label. After application of the subject-preposition deletion rule, the form of the sentence is that represented in 58.

(58)

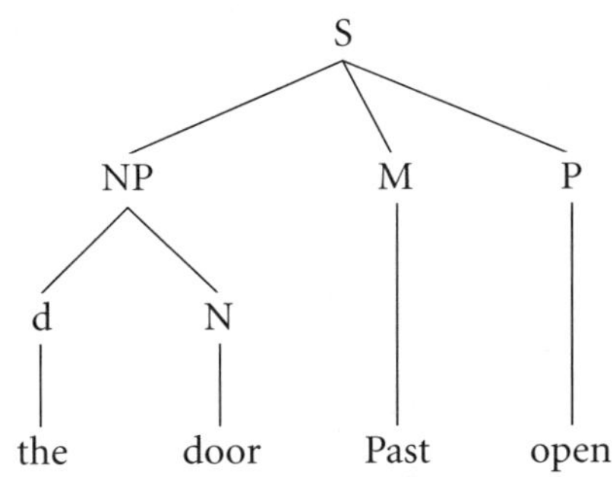

The final surface form, shown in 59, results from incorporation of the tense into the verb.

(59)

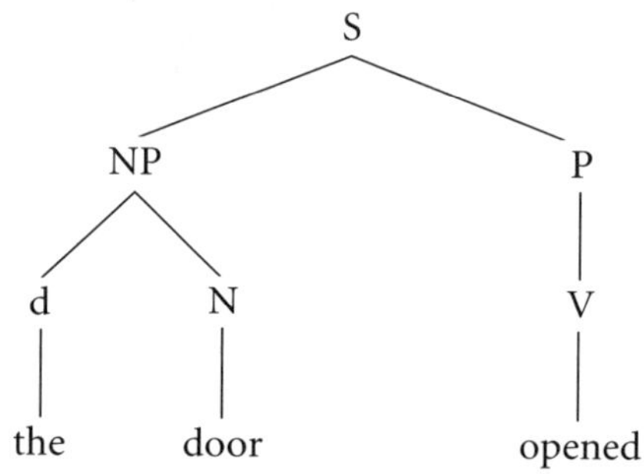

For a base configuration containing an A, a distinction must be made between the 'normal' and the 'nonnormal'[42] choice of subjects. The choice of the A as the subject, in accordance with the rule proposed in 55 above, requires no modification of the verb. The changes from 60 to 61 represent subject-fronting, those from 61 to 60 show subject-preposition deletion, and those from 62 to 63 indicate the effect of a third rule, object-preposition deletion.[43] The eventual surface structure of the sentence whose deep structure is 60 to 64.

42. The choice of terms is not to be taken seriously.

43. Verbs are categorized according to whether they delete the preposition of the following case category, that is, whether they 'take on' a direct object. The object-preposition deleting property of a verb may be modified by a transformation.

(60)

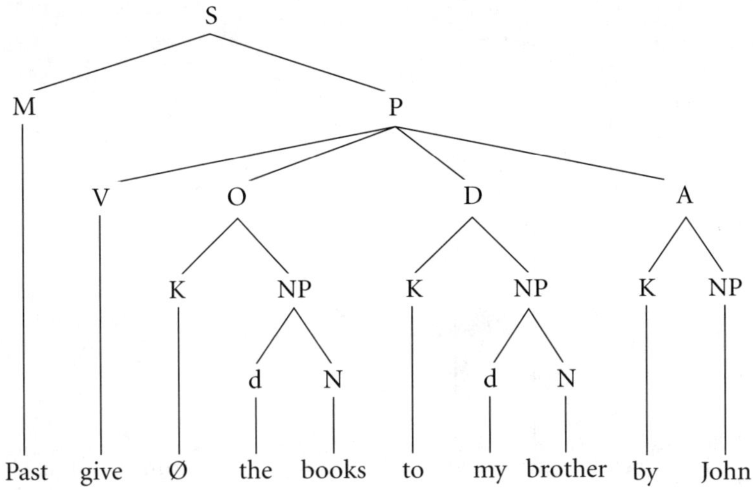
S
M P
V O D A
K NP K NP K NP
d N d N
Past give Ø the books to my brother by John

(61)

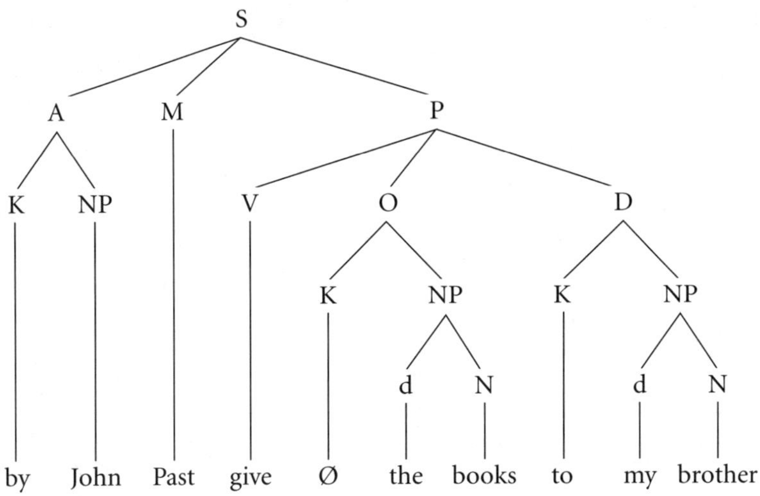
S
A M P
K NP V O D
K NP K NP
d N d N
by John Past give Ø the books to my brother

(62)

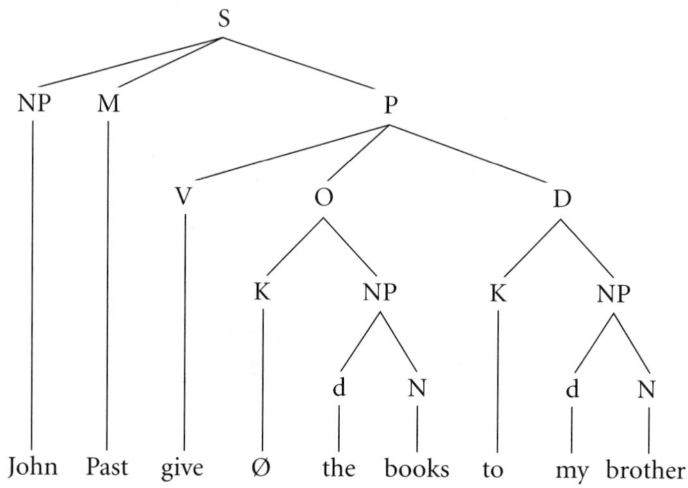
S
NP
M
P
V
O
D
K
NP
K
NP
d
N
d
N
John
Past
give
Ø
the
books
to
my
brother

(63)

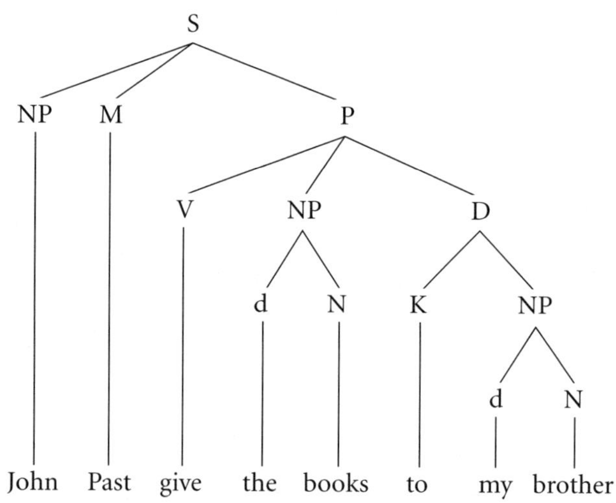
S
NP
M
P
V
NP
D
d
N
K
NP
d
N
John
Past
give
the
books
to
my
brother

(64)

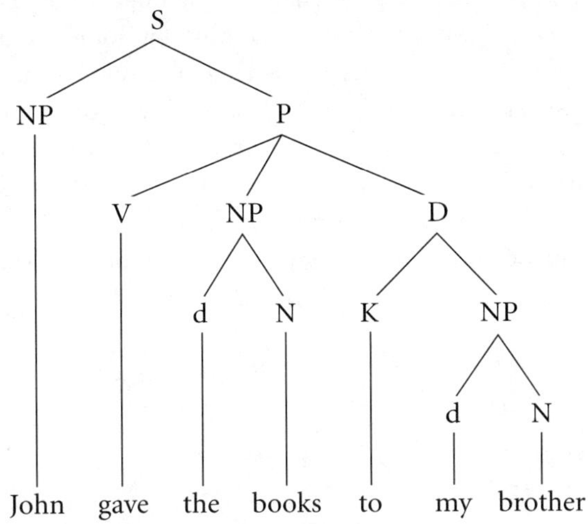

If it is noted that the verb *give* is one which, with A as subject, allows either O or D to become the direct object, an alternative surface form for 60 is 65 (assuming that case-label deletion occurs when zero K elements are 'deleted').

(65)

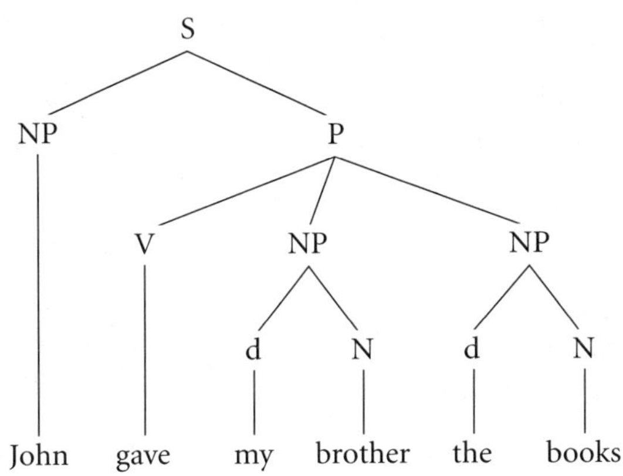

The 'normal' choice of subject for sentences containing an A, as stated in generalization 54 (which is a generalization for English), is the A. The verb *give* also allows either O or D to appear as subject as long as this 'nonnormal' choice is 'registered' in the V. This 'registering' of a 'nonnormal' subject takes place via the association of the feature [+passive] with the V. This feature has three effects: the V loses its object-preposition deletion property, it loses its ability to absorb the tense (requiring the automatic insertion of a *be* in the M constituent), and it must now be filled by a special 'passive' form (that is, *given*). The sequence 66 to 69 develops the choice of O as subject, and the sequence 70 to 74 shows the result of choosing D as subject.

(66)

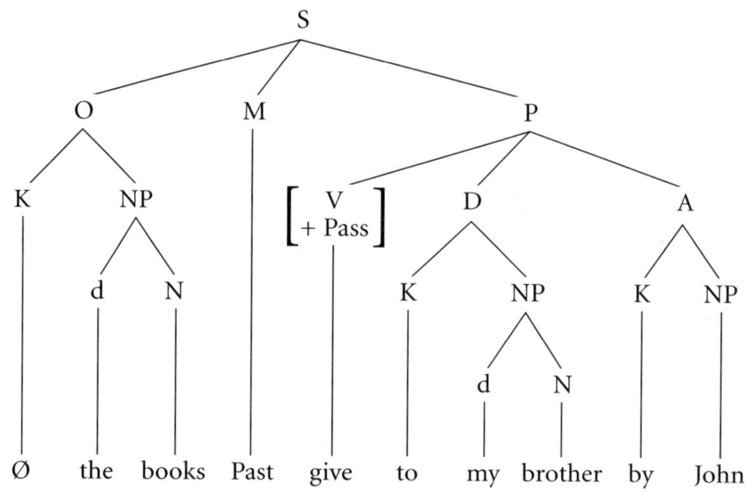

(67)

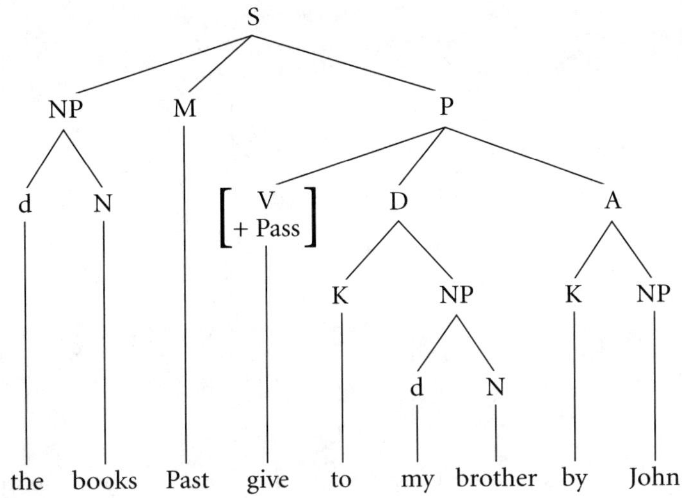
S
NP M P
d N
[V + Pass] D A
K NP K NP
d N
the books Past give to my brother by John

(68)

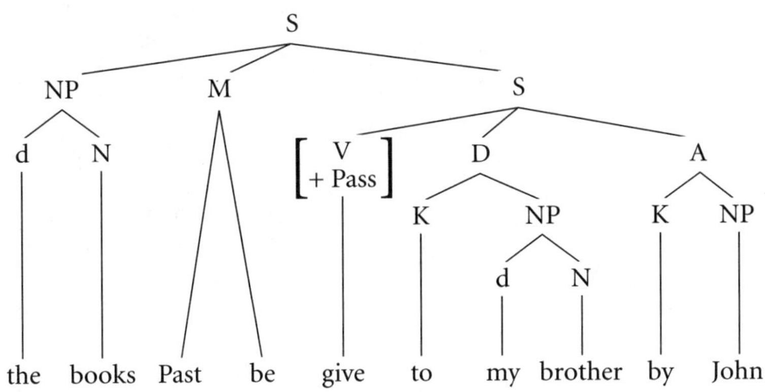
S
NP M S
d N
[V + Pass] D A
K NP K NP
d N
the books Past be give to my brother by John

(69)

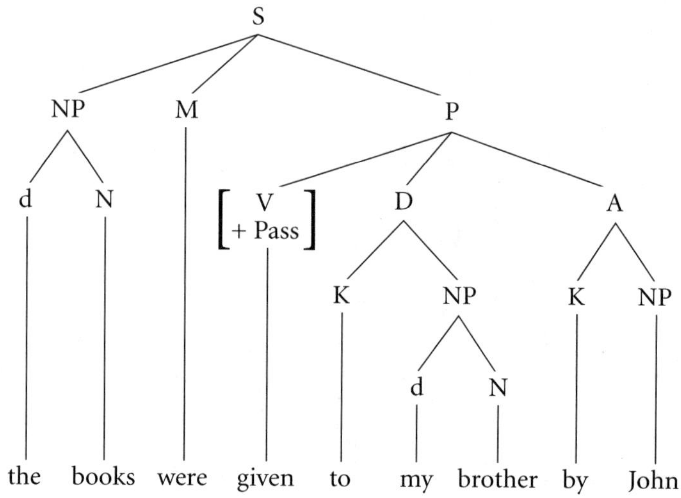
S
NP M P
d N [V + Pass] D A
K NP K NP
d N
the books were given to my brother by John

(70)

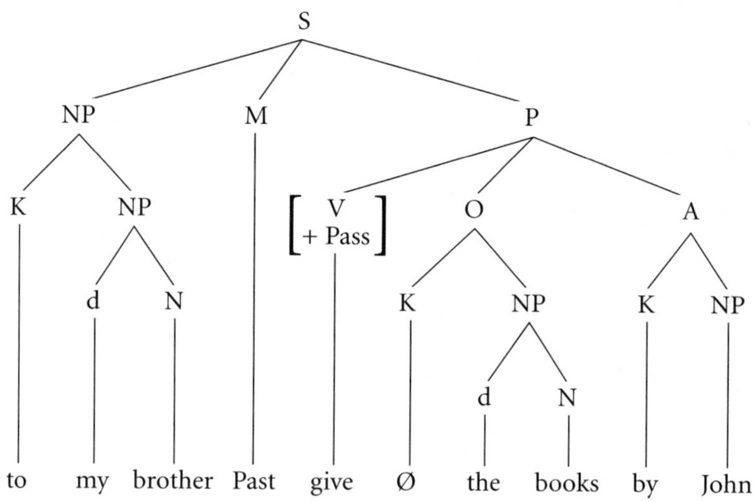
S
NP M P
K NP [V + Pass] O A
d N K NP K NP
d N
to my brother Past give Ø the books by John

(71)

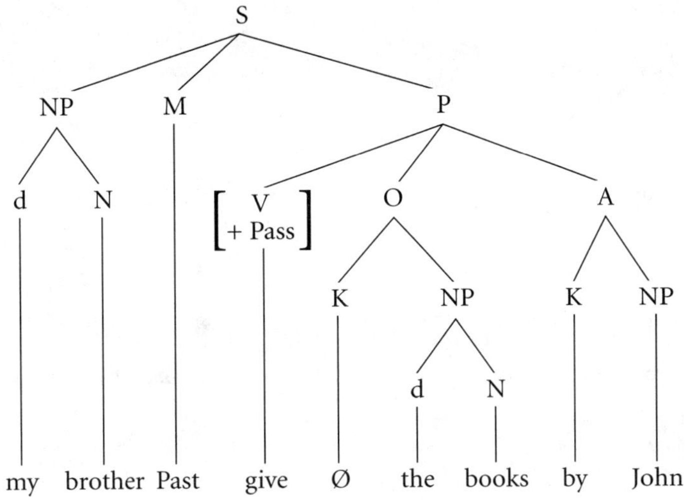
S
NP
M
P
d
N
V
+ Pass
O
A
K
NP
K
NP
d
N
my
brother
Past
give
Ø
the
books
by
John

(72)

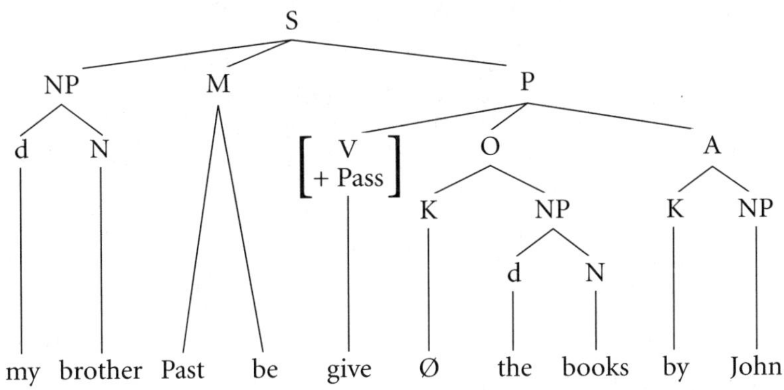
S
NP
M
P
d
N
V
+ Pass
O
A
K
NP
K
NP
d
N
my
brother
Past
be
give
Ø
the
books
by
John

(73)

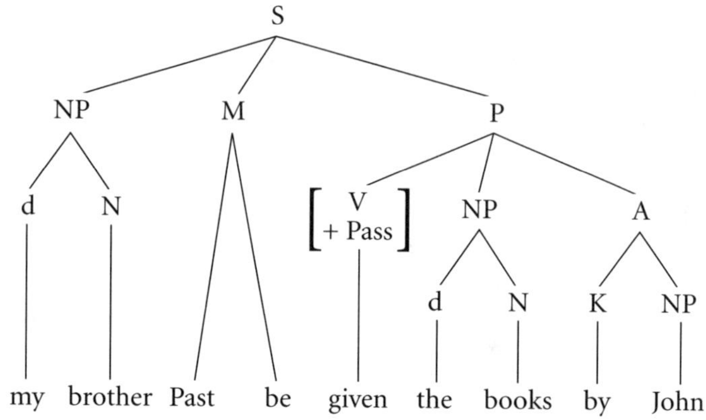

(74)

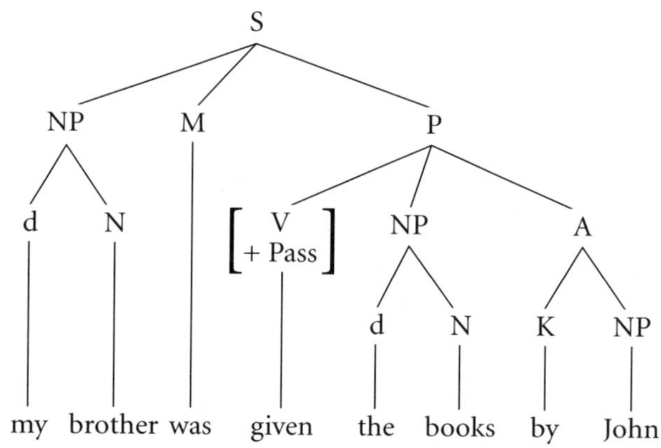

We have seen that where there is only one case category, its NP must serve as the surface subject. Examples 60 to 74 have shown ways of dealing with sentences containing more than one case category where one designated case could provide the subject without effecting any change in the V, or others could do so as long as a 'record' of this decision was attached to the V.

For many of the verbs which 'take' more than one case category, the one which contributes the subject is indicated by the verb itself. Of the verbs which are accepted into the frame [____O + D], *please, belong, interesting,* and others choose O as subject, and *like, want, think,* as well as others, choose D.[44]

Sometimes subjects are created not by moving one of the case elements into the 'subject' position, but by *copying* a particular element into that position. This seems to be a consequence of the positional treatment of subjects in English and to be related to the use of purely formal subjects.[45]

Copying for pro replacement can be illustrated with *that* clauses. The 'verb' *true* occurs in the frame [____S], that is, as in Configuration 75.

(75)

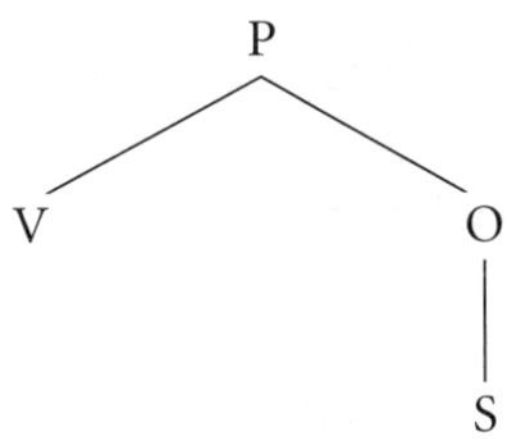

Since there is only one case element, it is obligatorily the subject. The context requires that the complementizer *that* be provided for the embedded sentence. By subject copying, 77 is derived from 76.

44. As mentioned above, by regarding the differences here as representing no more than idiosyncratic facts about the syntactic properties of these verbs, we can accept historical changes like those with *like, want,* and *think* from verbs of the type which choose O to verbs of the type which choose D to be merely a matter of detail in the subject-selection processes in our language. In other words, we do not need to agree with Jespersen when he describes the change in English from the use of expressions of the type 'him like oysters' to those of the type 'he likes oysters' as reflecting a change in the 'meaning' of the verb *like* from something like 'to be agreeable to' to something like 'to take pleasure in' (Jespersen 1924, p. 160). The change seems merely to be a result of the inter-influencing of the two surface processes of choosing the first word and establishing verbal concord.

45. From the fact that there may be only one case in a simple sentence, it becomes possible to allow all subjects to be formed by a copying transformation. Sentences with two copies of the same NP in the same case undergo one of a number of changes: the second copy is either deleted or replaced by a pro form, or the first copy is replaced by a pro form.

(76)

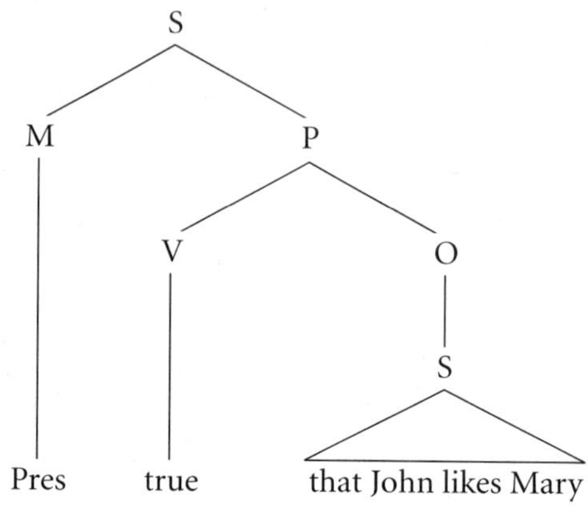

(77)

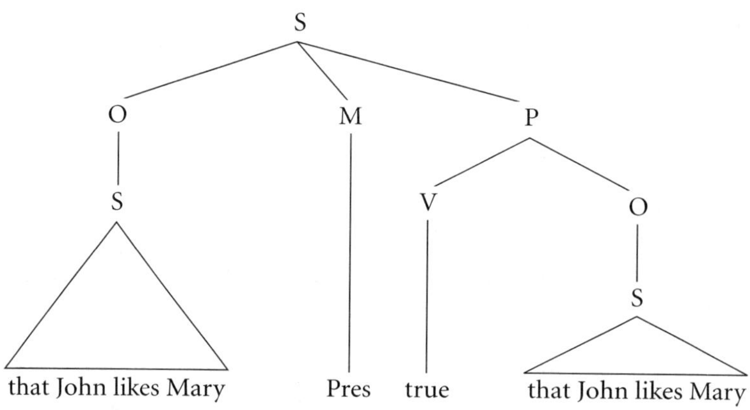

The structure of 77 undergoes either second copy deletion, yielding 78, or first copy pro replacement, giving us 79.

(78)

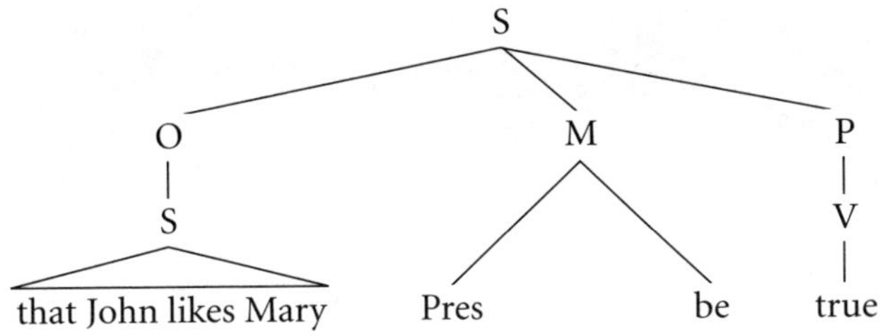

(79)

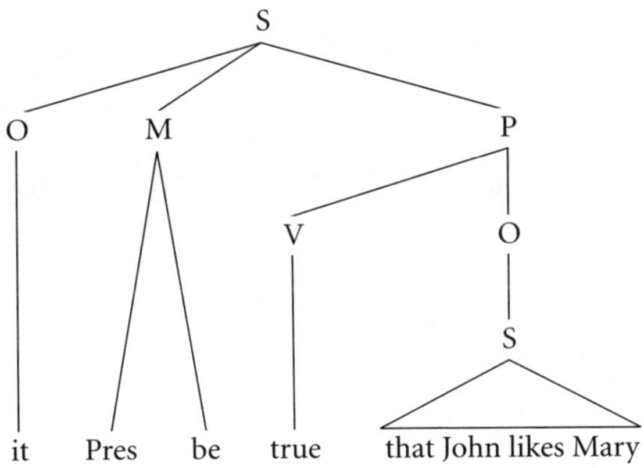

Verbs expressing meteorological conditions have the frame feature +[____L]. Choosing *hot* in that frame, we can construct the sentence whose deep structure is represented by 80. From 80 we get, by subject copying, item 81. By second copy deletion (and subject-preposition deletion) item 81 becomes 82; on the other hand, if the first copy is replaced by its pro form (in this context, *it*), the resulting sentence is 83.[46]

---

46. It is likely that the correct analysis of subject copying is a little different from this. There is considerable evidence that when the first copy is replaced by its pro form, the second copy is actually outside of P, that is, that it is 'extraposed' in the sense of Rosenbaum. If this is true, then since the sentences having undergone extraposition must be created in two steps anyway, it is likely that the sentential subjects are formed in the usual way–not by copying–and that they are later extraposed, leaving behind, in the subject position, a 'trace' in the form of expletive *it.*

The examples and the analysis of meteorological verbs are adapted from Langendoen (1966).

(80)

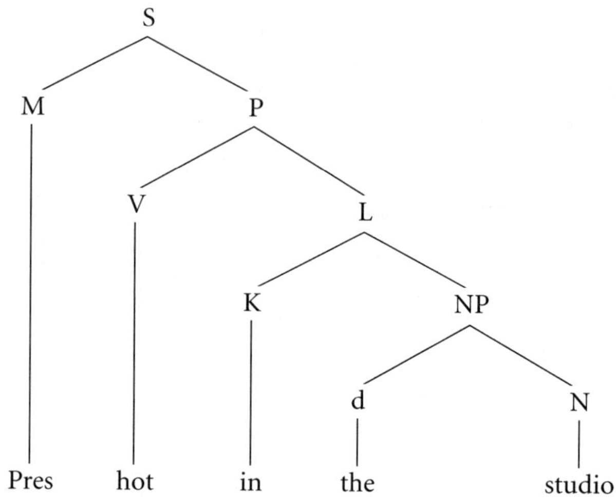
S
M
P
V
L
K
NP
d
N
Pres
hot
in
the
studio

(81)

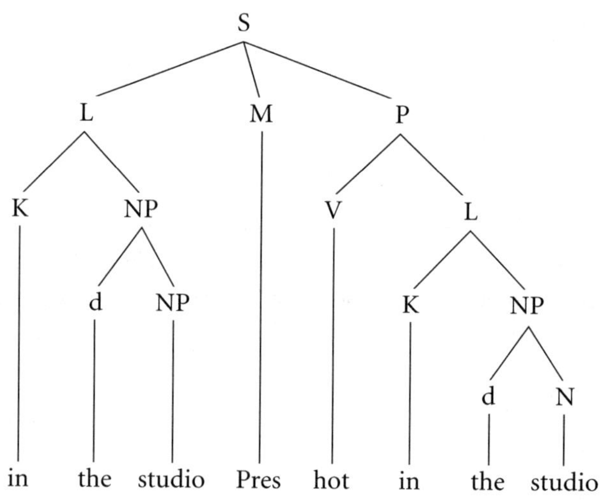
S
L
M
P
K
NP
V
L
d
NP
K
NP
d
N
in
the
studio
Pres
hot
in
the
studio

(82)

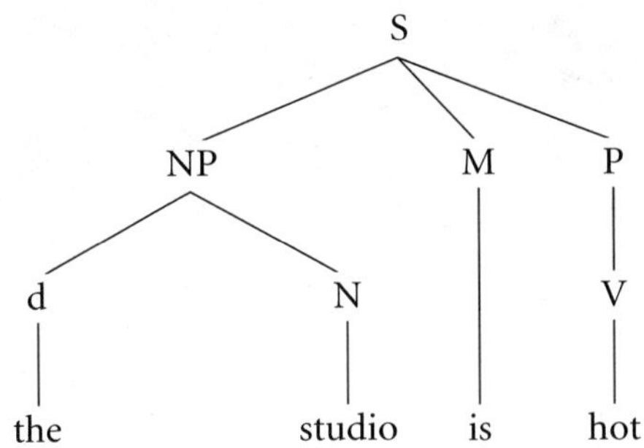

(83)

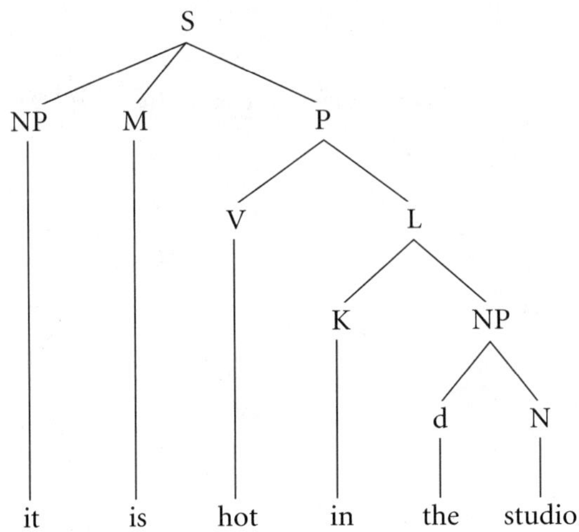

Under certain conditions, a first copy L may be replaced by an expletive *there*. The case frame [____O + L] may be filled by a blank verb (that is, zero). This situation (of verbless sentences) may call for the introduction of the element *be* into the M constituent, which is a process we have already seen to be necessary for verbs which are adjectives as well as for verbs which have been modified through addition of the feature [+passive]. For verbless sentences of the type [____O + L], the 'normal' subject choice is O. Thus from 84 we get 85, and eventually 86.

(84)

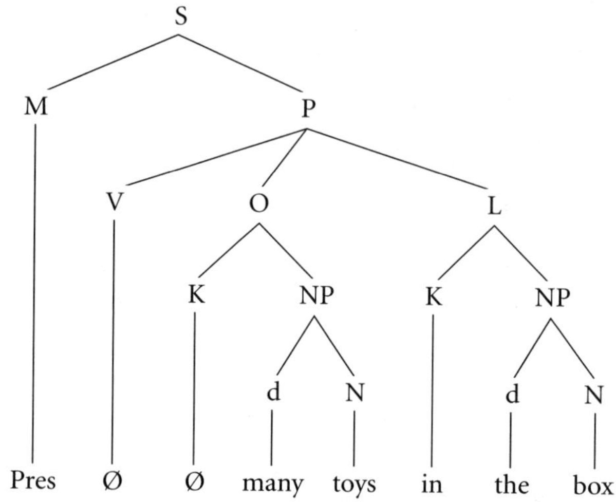
S
M
P
V
O
L
K
NP
K
NP
d
N
d
N
Pres
Ø
Ø
many
toys
in
the
box

(85)

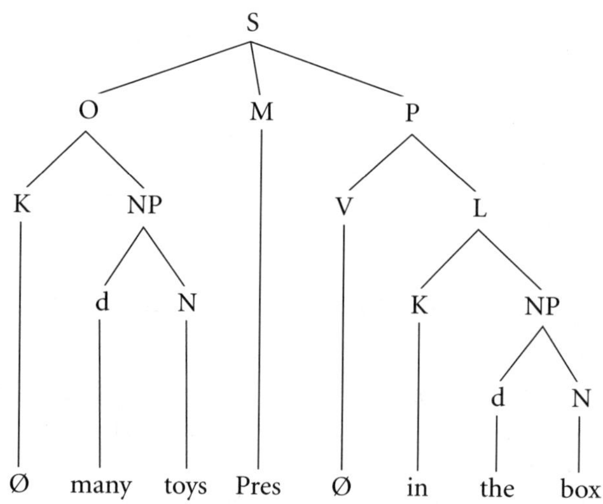
S
O
M
P
K
NP
V
L
d
N
K
NP
d
N
Ø
many
toys
Pres
Ø
in
the
box

(86)

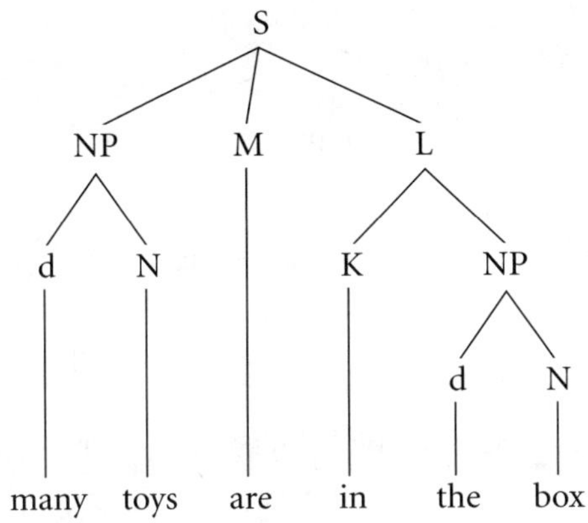

An alternative subject choice, through subject copying, is the L. Thus from 84 we might get 87.

(87)

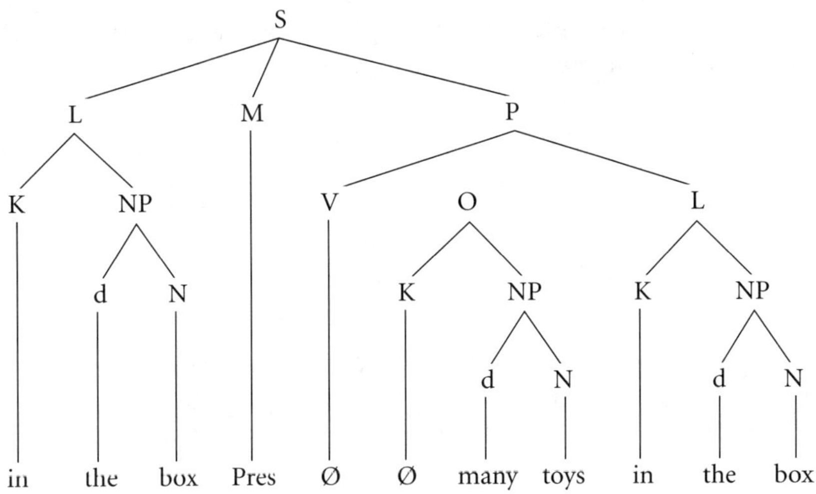

The pro form for L in verbless sentences is expletive (unstressed) *there*. The result of modifying 87 by pro replacement of the subject L is 88; extraposition of the second copy L, as suggested in footnote 45, has been carried out in 88.

(88)

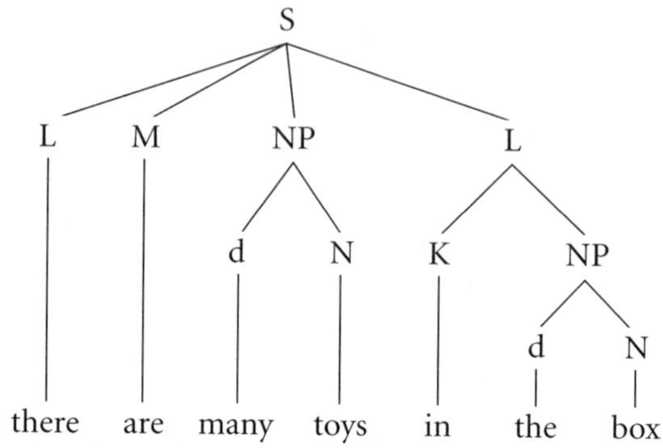

An alternative to replacing the first copy L by expletive *there* is to retain the L NP as subject. This decision requires the regular pronominalization of the repeated NP. It further requires modification of the verb: the hitherto empty V position is filled with the function verb *have.*[47] Since *have* is a V, it is capable of absorbing the tense, making the addition of *be* to M no longer necessary. The result of choosing the first L as subject results, through subject-preposition deletion, *have* insertion, object-preposition deletion, repeated NP pronominalization, and tense affixation, in 89.

(89)

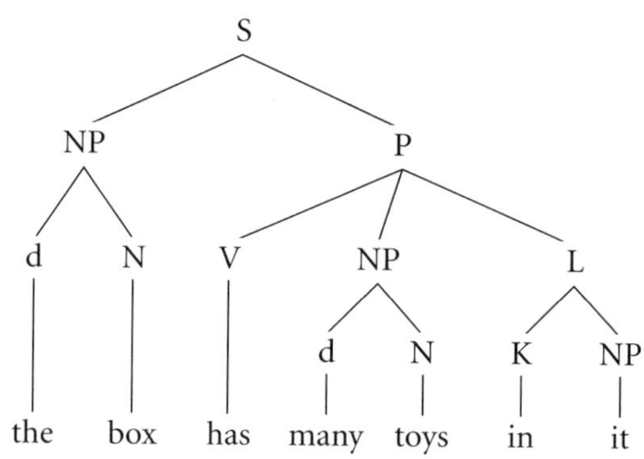

47. For a recent argument on the transformational introduction of *be* and *have* in all of their occurrences, see Bach (1967). For a more adequate treatment of existential sentences than I have presented, see especially Lee (1967).

The general position I am taking on the verb *have* is that in verbless sentences (that is, when the V constituent is present but lexically empty) *have* is obligatorily inserted just in case the subject is an NP which is not from the case O. The most obvious case is that of the empty verb in the frame [______O + D], a context which in English requires D to be the subject, resulting in the typical *have* sentences. Other languages, for example, French, seem to have contexts in which the subject choice is optional–situations where *X a Y* is in a paraphrase relation with *Y est à X*. Other languages, for example, Estonian, do not have anything equivalent to the verb *have*.[48]

Some languages have subjectivalization processes; and, as I have suggested for English, there seems to be an analogous objectivalization process which has the superficial effect of bringing a particular nominal element into closer association with the verb.

The formal rather than purely notional character of the direct object was noticed by Jespersen. His examples (1924, p. 162) show intra-language paraphrase relations like that between 90 and 91, and cross-language differences like that between 92 and 93.

(90) present something to a person
(91) present a person with something
(92) furnish someone with something
(93) *fournir quelque chose à quelqu'un*

When such phenomena were examined by Hall, she took one form as basic, the other as derived. 'Derived subjects', in her analysis, are possible just in case there is no 'deep subject'; 'derived objects', on the other hand, have the effect of displacing the original deep-structure object and attaching a *with* to it. Her examples include 94–95 and 96–97.

(94) John smeared paint on the wall.
(95) John smeared the wall with paint.
(96) John planted peas and corn in his garden.
(97) John planted his garden with peas and corn.

---

48. Another situation for introducing *have* to account for connections between such pairs of sentences as i and ii is discussed below in the section on inalienable possession.

(i) My knee is sore.
(ii) I have a sore knee.

Hall provides rules which move the locative element (*the wall* or *his garden* of 94 and 96 respectively) into the direct object position by a transformation which also attaches *with* to the former direct object.

From the point of view taken here, it would be just as easy to say that both *on the wall* and *with paint* were initially provided with prepositions (as L and I case elements), the verb *smear* having the property that whichever of these elements is chosen as 'direct object' must fall next to it and must lose its preposition. (In other languages, the process might be expressed as converting an original case specification to 'accusative'.)[49]

Subjectivalization, where it occurs, results in a neutralization of underlying case distinctions to a single form, usually called the 'nominative'. Objectivalization, where it occurs, neutralizes case distinctions to a single form which, where it is distinct from the form assigned to subjects, is traditionally termed 'accusative'. A third process which has the effect of effacing deep-structure case distinctions is the formation of nominals from sentences. The case modifications under nominalization transformations usually involve what is called the 'genitive'.

---

49. There are semantic difficulties in treating subject and object transformationally, in the sense that different choices are often accompanied by semantic differences of one sort or another. These differences are more on the order of 'focusing'–to be as vague as possible–than anything else, and do not seem to require positing 'subjects' and 'objects' in the deep structure. The 'focusing' difference may be extremely slight, as in the pairs i–ii and iii–iv, or it may have somewhat more 'cognitive content' as in the pairs v–vi and vii–viii.

(i) Mary has the children with her.
(ii) The children are with Mary.
(iii) He blamed the accident on John.
(iv) He blamed John for the accident.
(v) Bees are swarming in the garden.
(vi) The garden is swarming with bees.
(vii) He sprayed paint on the wall.
(viii) He sprayed the wall with paint.

Sentence vi seems to suggest, while v does not, that the whole garden has bees in it everywhere; and viii suggests, while vii does not, that the entire wall got covered with paint.

To the extent that other grammars make use of derived subjects and derived objects–which is the only alternative, within subject/object grammars, to treating verbs like *spray, blame, open, break* as involving elaborate and unexplained examples of homonymy–the semantic difficulties are just as great for them as they are for case grammar. Since the 'semantic effect' of the transformations in question is so different in kind from the semantic role of the case relations themselves, and since the latter are not affected by these processes, I am inclined to tolerate the reintroduction into grammatical theory of transformations which have semantic import (of this highly restricted kind).

The brief mention above of situations in which there was an S embedded to the case category O suggested the ways in which case grammar must deal with verb and adjective complementation. A second source of embedded sentences is within the NP itself. The rule for NP may be stated as 98.

(98) NP → N (S)

Where the N is an ordinary lexical item and the adjunct S contains a coreferential copy of the same N, the result is an NP consisting of a noun modified by a relative clause. One of the most obvious sources of 'genitive' is from relative clauses built on sentences which, by themselves, would have assumed the form *X has Y.* The N in the modified NP is the same as the N contained in the D of the adjunct sentence, and the V is empty. Thus, from 99 we get 100 by deleting the repeated noun, the tense, and the 'empty' verb and reattaching the D to the dominating NP.

(99)

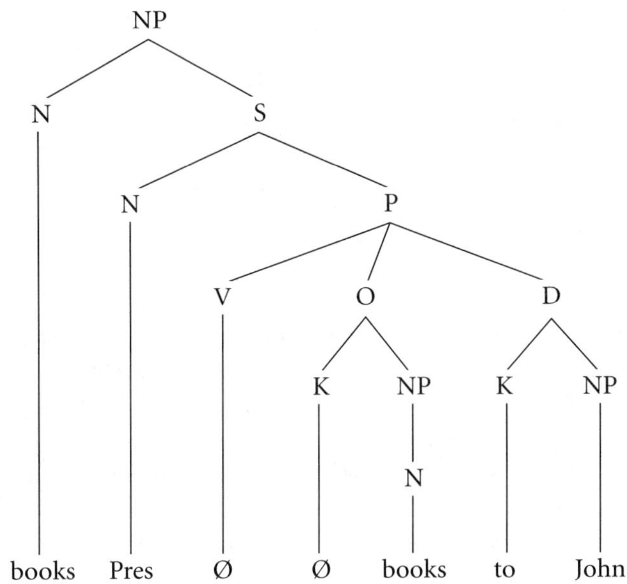

(100)

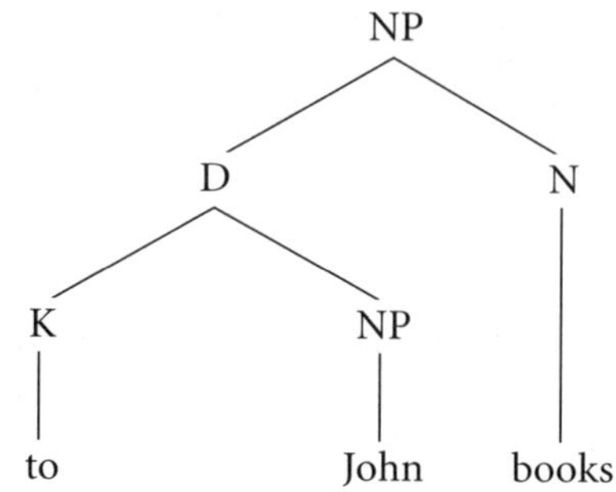

A D subjoined to an NP has its case marker modified–in this case to the sibilant suffix. Note 101.

(101)

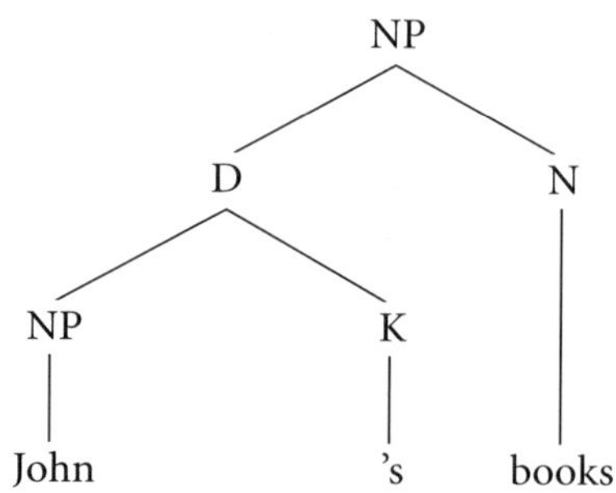

The 'true possessive' construction–resulting either in an NP of the form *X's Y* or *Y of X* in English–has as its source a sentence which by itself would have the form *X has Y.* The fact that in some languages there are instances of adnominal D not modified to the 'genitive' (*dem Vater sein Haus*, 'dative of possession') supports the view that conversion to genitive is a matter of the surface structure.

The interpretation of deverbal nouns which seems most satisfactory to me is that, except for the purely productive cases, the derivation of a noun from a verb is a matter of historical, not synchronic, fact. The synchronic reality is expressed by indicating that a given noun has a particular kind of relationship to a specific verb (or set of verbs), and that some of these nouns may, others must, appear in the NP frame [_____S].

That is, instead of having a synchronic process for producing such words as Latin *amor* from its associated verb, what is needed is the classification of such a word as an abstract noun having a particular kind of relationship with the verb *amo.*[50] Nouns having this kind of special relationship to specific verbs can take part in a process which introduces into the NP elements which 'originally' depended on the associated verb. The processes in question frequently have the effect of converting the form of the subsidiary NPs to the genitive.[51] Thus the noun *amor* when qualified by a sentence of the form *deus amat ...* yields *amor dei*; when it is qualified by a sentence of the form *deum amat ...* the result is again *amor dei.* The D and O forms, in other words, are equally reduced to the genitive, and when only one noun is involved, potential ambiguities result.[52]

---

50. This treatment allows for the inclusion of nouns which lack etymological connections with their related verbs. We might wish to indicate for *book* a connection of the intended kind with the verb *write*, thus accounting for the ambiguity of *your book* between 'the book which you own' (ordinary relative clause modification) and 'the book which you wrote'.

51. Exactly what universal constraints there are, if any, on the element to be converted to genitive is not at all clear. It appears that if there is only one element that shows up in the NP, it frequently takes the genitive form. Compare the ambiguous sentence i with sentences ii and iii.

(i) My instructions were impossible to carry out
 (a) so I quit.
 (b) so he quit.
(ii) My instructions to you are to go there.
(iii) *My your instructions are to go there.

In English it appears that if the conditions which allow the formation of the *of* genitive and the *s* genitive are satisfied by two different NPs in the associated sentence, multiple genitive constructions become possible, as in the following example borrowed from Jespersen.

(iv) Gainesborough's portrait of the duchess of Devonshire.

Japanese allows conversion to genitive in true relative clauses, as well as in the reduced relative clauses. A paraphrase of v is vi; *no* is the postposition most closely associated with functions which we would call 'genitive'.

(v) *Boku ga yonda zasshi.* 'I + subject + read – past + magazine' 'the magazines I read'
(vi) *Boku no yonda zasshi.*

52. Jespersen's suggestion that the ambiguity of *amor dei* is in the verb rather than the noun–the noun unambiguously identifying the subject, the verb being ambiguously either active or passive–must be understood as the hypothesis that only those NP constituents which are capable of conversion to surface subjects (with a given verb) may appear under genitive modification as modifiers of the deverbal noun. For English this may well be true. (Jespersen 1924, p. 170.)

4. Some Remarks on Language Typology

The view of universal grammar which is emerging is something like this: In their deep structure, the propositional nucleus of sentences in all languages consists of a V and one or more NPs, each having a separate case relationship to the P (and hence to the V). The most straightforward deep-structure commonalities between languages are to be sought at this 'deepest' level.

The lexical insertion rule for verbs is sensitive to the particular array of cases in the P. Since no distinction is needed between 'strict subcategorization features' and at least the highest level of 'selectional features' (because redundancy relations exist between cases and some lexical features, and because there is no 'subject' outside of a 'VP' whose features need to be dealt with separately), the lexical insertion rule for verbs can be a strictly local transformation which responds to nothing more than the cases which are co-constituents of V (with the exception, as noted above, that it must be known whether the O element is an NP or an S).

The criteria for typological classification that have suggested themselves so far in this study are these:

- I. the presence or absence of modifications on the NPs as determined by the deep-case categories
    - A. the nature of such modification (prepositional, affixal, or other)
    - B. the conditions for the choice of particular case forms (which, when stated in their simplest form, constitute what is usually formulated as the 'case system' of the language)
- II. the presence or absence of concordial modifications of the verb
    - A. the nature of the concord (number agreement, incorporation of 'traces' of case categories, feature changes on V)
    - B. the relation to subject selection (topicalization)
- III. the nature of anaphoric processes
    - A. type of process (replacement by pro form, deletion, de-stressing, replacement by unstressed variants, or other)
    - B. conditions of application
- IV. topicalization processes (where 'subject selection' may be thought of as a special case of topicalization)
    - A. formal processes (fronting, modifying the case form, or other)
    - B. the variety of topicalization processes in the same language
- V. word order possibilities
    - A. factors determining 'neutral' word order (nature of case

categories, 'ranking' of noun classes, topic selection, or other)

B. conditions determining or constraining stylistic variations on word order

It is important to realize that all of these typological criteria are based on superficial processes, and that there are no particularly good reasons for believing a priori that there will be much coincidence in the ways in which the different criteria sort out the world's languages.

4.1. *The Bases for Determining Case Forms*

The forms of the NPs in a P are determined on the basis of a variety of factors, one of which is the case category of the NP. Thus an NP under an I (that is, an instrumental noun) is assigned a particular form depending in whole or in part on the fact that it is under I.

Surface case forms of NPs are most elaborately developed in the personal pronouns. The study of the 'case' aspects of pronoun systems reveals a great deal about the variety of relationships that can hold between deep and surface cases.

Sapir's typological distinctions for Amerindian pronominal systems (1917) can be expressed in case grammar terms quite simply. If we ignore whatever complications may exist in 'passive' constructions, and if we ignore all deep-structure cases except A and O, we can imagine sentences of the following three types given in their underlying propositional form:

(a) V + A   intransitive sentences with active 'subjects'
(b) V + O + A   transitive sentences with agents
(c) V + O   intransitive sentences with inactive 'subjects'

Since the V element is constant to the formulas, we can represent these three sentence types by presenting the case frames in three lines, as follows:

(102)

	A
O	A
O	

According to Sapir, then, there are languages which, like Yana, have only one form for pronouns in all four of these positions.

(103)

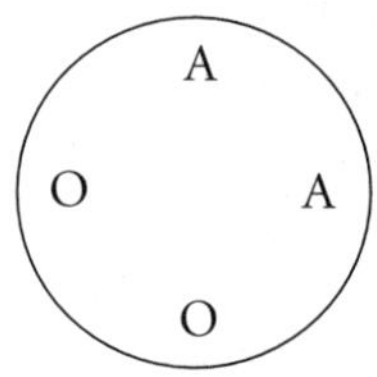

There are languages like Paiute that have a separate form for the O element in the transitive sentence, all others being the same. The two forms are traditionally called 'nominative' and 'accusative'.

(104)

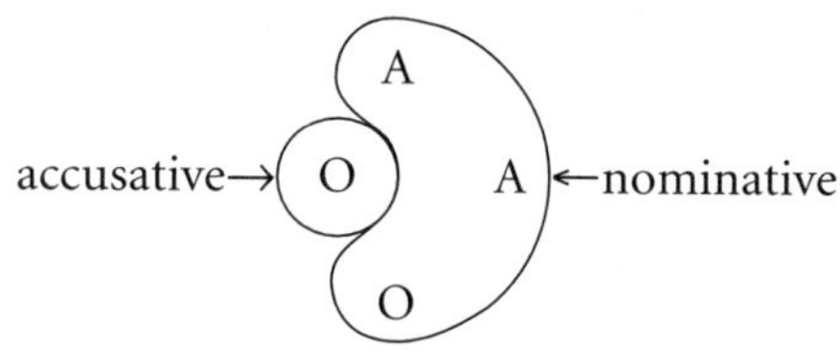

There are languages like Chinook which give one form to the A of transitive sentences and another to the remaining cases. The terms 'ergative' and 'nominative' are often given to a distinction made in this way.

(105)

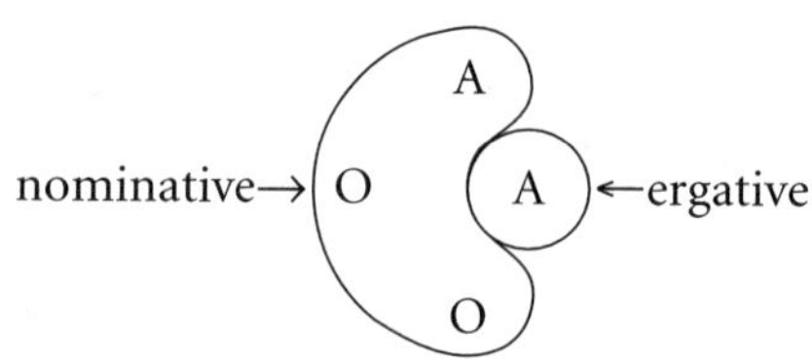

There are languages like Dakota which have separate forms for A and O; here the terms are usually 'active' and 'inactive'.

(106)

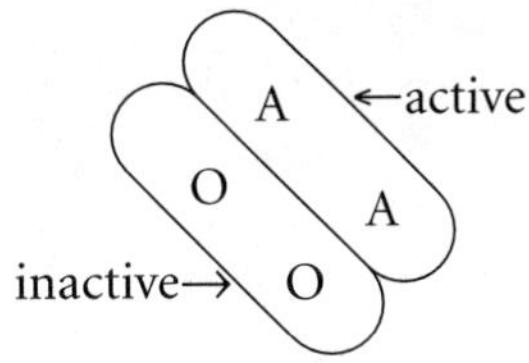

And, lastly, there is the situation found in Takelma, which has one form for the pronominal NP of intransitive sentences, and two separate forms for the A and O of transitive sentences. Thus:

(107)

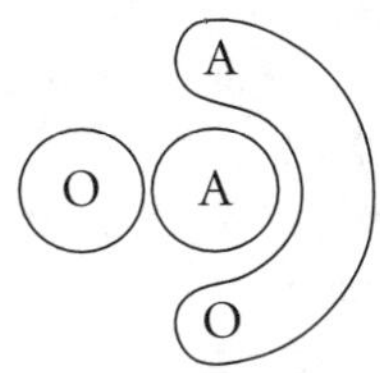

What these observations are intended to suggest is merely that if I correctly understand Sapir's analysis of the pronominal systems of these languages, then the case concepts I have been discussing, together with the notion of clause types which various arrays of them define, provide the categorial and configurational information for determining the surface distinctions that are found in these languages.

4.2. *Verbal Concord*

There are these various ways, and possibly more, in which cases and case environments are involved in determining the case forms of NPs within a P. An additional factor is whether the given NP has been chosen as subject in languages having subjectivalization processes. Choosing subjects or topics is related to another aspect of the superficial structure of sentences, and that is verbal concord.

The choice of subject in English always has the effect of determining number concord (on those verbal and auxiliary elements capable of reflecting number concord). Apart from number concord, the choice of subject might involve modification of the verb to its passive form, or introduction of the verb *have.*

The information 'registered' in the V may have only to do with the choice of subject, as in English, or it may be more elaborate. Languages which 'incorporate' pronominal affixes into the V may do so for more than one NP at a time; or noun stems themselves from particular cases may be incorporated into the verbal expression.[53]

The subject selection rules discussed for English may be compared with the topicalization processes that are described for Philippine languages. The situation for Maranao has recently been described by McKaughan. One NP is chosen as topic for every sentence, and this choice is recorded in the following way: its original case preposition is replaced by *so*, and an affix is inserted into the V which indicates the case category of the chosen NP. There is apparently considerable freedom in the choice of topic. To take the verb meaning 'to butcher' (/*sombali?*/), we find that when the topic noun is an original I, the verb takes on the prefix /*i-*/, as shown in 109; and when the topic is an original B, the suffix /*-an*/ is added to the V, as seen in 110.

(108) som*ombali?* so *mama? sa karabao*
'The man butchers the carabao.'

(109) i*sombali? o mama?* so *gelat ko karabao*
'It is with the knife that the man butchers the carabao.'

(110) *sombali?* an *o mama?* so *major sa karabao*
'It is for the mayor that the man butchers the carabao.'[54]

The choice of sentence subjects, or 'topics', from particular cases appears to be the most satisfactory way of accounting for the many types of voice modifications of verbs such as those described as middle, pseudoreflexive, and so forth, in the Indo-European languages.

4.3. *Anaphoric Processes*

Anaphoric processes are best understood from the point of view of an extended concept of sentence conjunction. That is, every language has ways of simplifying sentences connected by conjunctions or subjunctions, and the processes used under these conditions seem to be exactly the same as those used in sentences connected in discourse. The grammarian's job, therefore, is to describe these processes as they work in sentences that are independently

53. Grammatical devices for providing concord of this type have been worked out for Mohawk by Paul M. Postal (see Postal, 1963).

54. McKaughan (1962). The examples and the description of the relationships are from McKaughan, but a great deal of guessing lies behind my interpretation.

intelligible, and then to assume that utterances in connected texts or conversations can best be understood from the point of view of a shared knowledge of the language's anaphoric processes on the part of speaker and hearer.[55] The fact that in these anaphoric or reduced forms English uses pro-replacement under conditions that would call for deletion in some other language may thus be seen as a superficial difference between the two languages.

The point is important–and it was mentioned above in connection with 'bad' reasons for rejecting the universality of the subject/predicate division–because the absence of subjects in the final surface forms of sentences in some languages is seen by many scholars as having great typological relevance. The optional absence of NP constituents in languages with person-marker incorporation (for example, Chinook) has led scholars to claim that such languages lack the nexus relations that Europeans understand as 'subject' and 'object' but have instead what are described as 'appositional' relations between NPs and Vs (see Sommerfelt, 1937). In languages without pronominal incorporation, a distinction is made by some scholars between true subject/predicate languages and those in which the so-called 'subject' is as much a 'complement' to the V as is the direct object or any of the various adverbial elements. To Martinet (1962a, pp. 61–62), a subject is different from a complement only if it is 'constitutive of the minimal utterance'–that is, only if it is obligatorily present in both full and anaphorically reduced utterances. In Japanese, the 'minimal utterance' lacks a subject, and hence, the argument goes, Japanese sentences lack the subject/predicate structure of sentences in our more familiar languages. To Martinet's disciple Saint-Jacques, this typological 'fact' about Japanese is regarded as excessively important. It is only by dint of considerable intellectual effort that the Westerner can achieve that liberation from familiar ways of thinking about language which is required for

---

55. In other words, the grammarian will describe the process by which i is converted to ii by noting the conditions under which repeated elements in conjoined sentences may undergo deletion and pro-replacement and under which conjoined sentences can have words like *too* and *either* added to them.

(i) Mary didn't want any candy and Mary didn't take any candy.
(ii) Mary didn't want any candy and she didn't take any either.

In contexts in which the information contained in the first conjunct of i is already understood by the addressee (by having just been spoken by him, for example), a speaker of English feels free to use the reduced form in iii.

(iii) She didn't take any, either.

There is no reason, it seems to me, to expect the grammar of a language to generate sentences like iii directly.

an understanding of the true character of Japanese. Or so Saint-Jacques (1966, p. 36) tells us. It seems to me that language typology offers enough genuine excitements to make it possible for us to give this one up. The intellectual achievement of which M. Saint-Jacques speaks is that of knowing that when there is an 'understood' NP to deal with, some people replace it by a pronoun, others get rid of it.

4.4. *Topicalization*

The fourth criterion has to do with topicalization processes, devices for isolating one constituent of a sentence as 'topic', of bringing one particular constituent of a sentence into some kind of 'focus'. Where topicalization is distinct from processes for 'emphasizing' a constituent, we have much the same thing as what I have been calling 'subjectivalization', but which I shall now begin calling 'primary topicalization'. Primary topicalization for English involves position and number concord; stylistic changes involving stress assignment, late word-order changes, and possibly the 'cleft-sentence construction' fall into what might be called 'secondary topicalization'. From what I understand of McKaughan's account (1962, p. 47), primary topicalization in Maranao involves replacement of the original preposition associated with a noun by *so* and introduction into the V of an associated case indicator, while secondary topicalization involves moving an NP to which *so* has been added to the front of the sentence. One might refer to Oertel's study of the disjunct use of cases in Brāhmanic prose as a study of secondary topicalization.[56] I would imagine that all languages possess some means of carrying out 'secondary topicalization', but it may be the case that some lack the process of 'primary topicalization' (subjectivalization').[57]

56. Oertel (1936) distinguishes 'pendent' uses of a disjunct case, where the 'topic' is in the 'nominative' even if its original role in the sentence was not that of subject (comparable, I assume, to *he* in 'he, I like him'), and 'proleptic' uses, where the topic retains the original case form, is moved to the front of the sentence, and may or may not be resumed (in the form of a demonstrative) in the remainder of the sentence (comparable to *him* in *him, I like (him)*).

57. Jeffrey Gruber's recent study (1967b) of topicalization in child language suggests that ontogenetically motivated (what I am calling 'secondary') topicalization precedes the use of formal subjects in English. It may be that when one device for topicalization becomes 'habitual', it freezes into a formal requirement and the language must then call on other processes for motivated topicalization.

Kenneth Hale (correspondence, 1967) reports that for Walbiri, an 'ergative' language of aboriginal Australia, there is apparently no 'subjectivalization' process, but any constituent may be repeated to the right of the proposition, the element inside the proposition being replaced by a pro form.

The notion 'subjectivalization' is useful only if there are sentences in a language which offer a choice of subject. Languages described as not having passives, or languages described as only capable of expressing transitive sentences passively, apparently lack the grammatical process of primary topicalization.

This question leads naturally to the problem of the so-called 'ergative' languages. Recall that in the accusative type of pronominal system, the pattern was

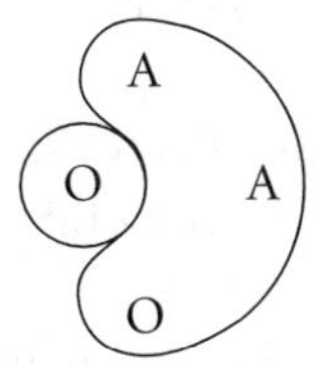

and that in the ergative type it was

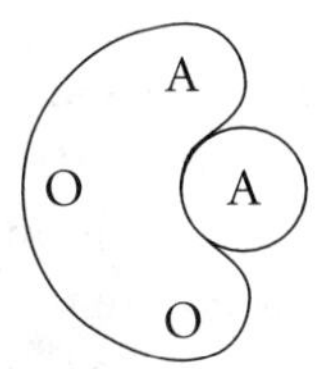

Now when languages of the accusative type have passive versions of sentences whose propositional form is [V O A], the case forms associated with the elements in the passive version are generally 'nominative' for the O and 'agentive' (realized as ablative, instrumental, or what have you, depending on the language) for the A. If passive sentences were introduced into our three-line diagrams and their active counterparts removed, we would get the pattern

(111)

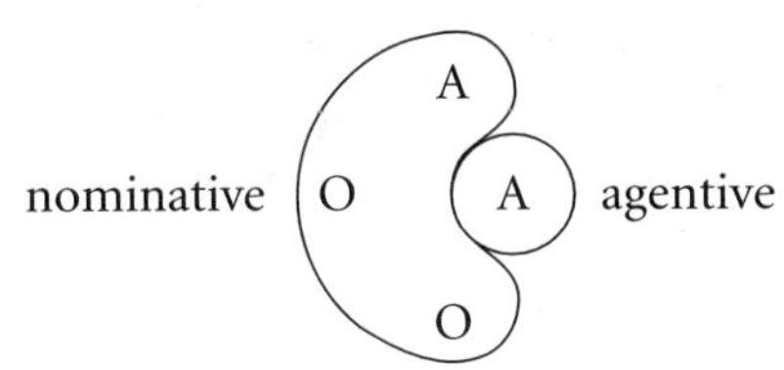

which is exactly like that for the regular assignment of cases in the ergative languages. This fact, plus the use of the term 'nominative' for subject-of-intransitive-*cum*-object-of-transitive in these languages, has led many scholars to identify the ergative case in ergative languages with the agentive case form found in passive sentences in accusative languages, and to conclude that the ergative languages are really 'passive' languages–languages in which transitive sentences can only be expressed passively.[58] For both of these systems, the case that has been given the name 'nominative' is frequently described as the 'subject' in a subject/predicate construction, and the 'ergative' element in the one instance and the 'accusative' element in the other are treated as verbal complements (see Trubetzkoy 1939). The difficulty of determining the 'subject' in ergative languages has been described by Martinet (1962b, pp. 78 f.): Some scholars identify as subject the word which would be the subject in a translation of the sentence into French–that is, the nominative in intransitive sentences and the ergative in transitive sentences. Others regard the nominative as the subject in all sentences, thus giving transitive sentences a 'passive' interpretation. Lafon gives up on transitive sentences–he uses the term 'subject' only for intransitive sentences, saying of transitive sentences that they have no subject.

Vaillant (1936, p. 93), on the other hand, spoke of the northern Caucasian languages as having three types of verbs: (*a*) true intransitives, with subjects in the 'nominative'; (*b*) 'operative pseudotransitives', with 'pseudosubject' in the 'ergative'; and (*c*) 'affective pseudotransitives', with 'pseudosubjects' in the 'dative' . It seems quite clear that what he is dealing with are sentences having Ps of the three types–[V O], [V O A], and [V O D]–where the surface cases for O, A, and D are 'nominative', 'ergative', and 'dative', respectively. It looks very much as if that is all there is to say. For my part I would much rather say of the ergative languages that they lack subjectivalization, than say either that all transitive sentences undergo obligatory passivization, or that some of their sentences contain true subjects while others do not.

The frequent claim that the ergative languages are more primitive than the accusative languages are (see Tesnière 1959, p. 112), together with the assumption that the ergative construction is really a passive construction, has led such scholars as Kuryłowicz, Schuchardt, and Uhlenbeck to assume that the passive construction represents a more primitive concept in the evolution of language than that of the active transitive construction. Evidence mustered

58. Note that even if there is a different form for the verb in [____O] and [____O + A] case frames, this cannot be interpreted as evidence of 'passivity'. As mentioned earlier, in languages not of the ergative type there may still be systematic variation of the same verb root depending on whether it is used transitively or intransitively.

for this position includes the signs that pre-Indo-European was of the ergative type, and the fact that some languages have 'invented' *have*-like verbs in relatively recent times. The invention of *have* made it possible to give active expression to certain tense or aspectual forms which had remained unaffected by the general change from passive to active expression (as is seen, for example, in the circa third century shift from expressions of the type *inimicus mihi occisus est* and *mihi illud factum est* to transitive expressions using *habeo: inimicum occisum habeo* and *habeo illud factum* (see van Ginneken, 1939, p. 86)).

It seems very unlikely to me that syntactic changes of the type known from the present state of our knowledge are really capable of showing an intellectual evolution of a type as potentially significant as whatever might be understood as the transition from an essentially passive to an essentially active point of view. The connection claimed by van Ginneken between ergativity and the 'feminine' character of cultures with ergative languages is another that should be questioned.[59]

4.5. *Word-Order Differences*

The fifth criterion suggested for a language typology is that of word order. The variables that determine or constrain the freedom of word order in the languages of the world are very likely to have many important connections with the case structure of sentences; but this is an area which I have not examined at all.

---

59. The following seems worth quoting in full (van Ginnken 1939, pp. 91 f):

> Nous sommes tous des hommes, et tous nous avons deux talents: les facultés plus actives de l'appétit et de la volonté, et les facultés plus passives des sensations et de l'appréhension; mais il est évident que les deux sexes de l'humanité montrent sous ce rapport une différence sensible.
>
> L'ethnologie moderne, qui a écarté définitivement comme insuffisante la doctrine du développement uniforme, nous apprend cependant que le progrès de l'humanité a balancé presque toujours entre les cultures plus féminines ou plus masculines, dites cultures matriarcales et patriarcales. Ce sont toujours les cultures matriarcales très prononcées qui, comme le basque, ont un verbe transitif de nature passive avec comme casus rectus un patiens et comme casus obliquus un agens; mais les cultures patriarcales, comme l'indoeuropéenne ont un verb transitif de nature active, animiste et magique, avec un sujet au casus rectus et un objet au casus obliquus. Chaque peuple a donc le verbe qu'il mérite.

5. The Grammar of Inalienable Possession

The preceding sections have contained an informal description of a syntactic model for language and a few demonstrations of the operations of this model of the sort that has come to be called 'restatement linguistics'. In the present section I shall attempt to show how a particular substantive modification of the rules will permit a uniform way of describing the interesting collection of grammatical facts associated with what is called 'inalienable possession'.

Every language, one can be sure, has nouns which express concepts that are inherently relational. Examples of inherently relational nouns in English are *side*, *daughter*, and *face*. One doesn't speak of a side, but of a side of something; one doesn't say of someone that she is a daughter, only that she is somebody's daughter; and although it is possible to speak of having seen a face, the word is typically used when referring to 'his face' or 'your face' or the like. The relational nouns most frequently discussed in the linguistic literature are names of body parts and names of kinsmen. My discussion here will concentrate on body parts.

5.1. *The Data*

5.1.1. Significant syntactic relationships exist between the dative and the genitive cases in all of the Indo-European languages; and in all but Armenian, according to Havers (1911, p. 317), the dative and the genitive case forms figure in paraphrase relationships of kinds that are highly comparable from language to language. The relationship is observed only when the associated noun is of a particular type. To take some of the modern German examples given by Havers, we observe that a paraphrase relation exists between 112 and 113 as well as between 114 and 115; but that of the two sentences 116 and 117, the latter is ungrammatical (as a paraphrase of 116).

(112) *Die Kugel durchbohrte dem Feind das Herz.*
(113) *Die Kugel durchbohrte das Herz des Feindes.*
(114) *Er hat mir die Hand verwundet.*
(115) *Er hat meine Hand verwundet.*
(116) *Der Vater baute seinem Sohn ein Haus.*
(117) **Der Vater baute ein Haus seines Sohnes.*

It should be noted that *Herz* and *Hand* are the names of body parts, while *Haus* is not.

5.1.2. There are cases like the above where a given language exhibits in itself the paraphrase relationship, and there are also cases where it appears that one language has chosen the dative expression, another the genitive. Notice the following sentences, also from Havers (1911, p. 1).

(118) My heart aches: *Mir blutet das Herz.*
(119) Tom's cheeks burned: *Tom brannten die Wangen.*
(120) She fell on her mother's neck: *Sie fiel ihrer Mutter um den Hals.*

5.1.3. There are adnominal (possessive) uses of dative constructions, particularly, it appears, when the possessive pronoun is also used with the possessed element. Here the most readily available examples are with kinship terms (Havers, 1911, p. 283).

(121) *Dem Kerl seine Mutter.*
(122) *Sa mère à lui.*

5.1.4. Many languages have separate possessive affixes for nouns that are obligatorily possessed (inalienables) and nouns that are optionally possessed (alienables). The difference in Fijian is apparently expressed by preposing the possessive morpheme to indicate alienable possession and suffixing it to indicate inalienable possession. Since the category 'inalienable' is a category of grammar rather than a property of real world objects (since, in other words, some objects grammatically classed as inalienable can in fact be separated from their 'owners'), the distinction can be seen most clearly if both methods of expression can be used with the same noun stem. Lévy-Bruhl (1916, p. 99) gives a persuasive example of this situation: Fijian *uluqu* means the head which is now firmly attached to my neck, while *kequ ulu*, also translatable as 'my head', would refer to the head which, say, I am about to eat.

Languages may have separate morphemes for indicating alienable and inalienable possession, and they may have further distinctions among these morphemes depending on the type of inalienable possession (as Nootka, for example, suffixes -*ʔat*- to nouns representing physically inseparable entities, for example, body parts, but uses other means for kinship terms), or they may merely have a class of nouns incapable of occurring as free forms–noun stems requiring affixation of possession indicators.[60]

60. This last situation is sometimes described by saying that nouns are 'inflected for person' (see Manessy 1964, p. 468).

The full variety of the treatment of inalienable possession in different Amerindian languages is catalogued in Sapir (1917a).

In all of these cases, it appears, the features in question are 'grammatical' rather than purely 'notional'. Discussions of inalienable possession almost always contain lists of nouns whose grammatical classification is the opposite of what one would notionally expect. Lévy-Bruhl (1916, p. 96) mentions a case where the word for 'left hand' functions as a body-part word grammatically, but the word for 'hand' does not. And Arapaho classifies 'louse' (or 'flea') among the inalienables (Salzmann 1965, p. 139), a situation that invites people who like to speculate on these things to propose something or other on the Arapaho conception of 'self'.

5.1.5. Milka Ivić (1962, 1964) has recently discussed many instances of what she calls 'non-omissible determiners'. Among the examples she cites are many that involve nouns of the type frequently included among the inalienables. The adjective cannot be deleted, for example, in the Serbo-Croatian expression in 123, for 124 is ungrammatical (1964, p. 477).

(123) *devojka crnih očiju*, 'the girl with black eyes'
(124) **devojka očiju*

What is misleading about her discussion, it seems to me, is the decision to associate with the adjective the 'category of nonomissibility'. It is as if we wished to say, for the English sentence 125, that there is something grammatically significant about the word *missing*, since its deletion results in sentence 126 which is somewhat different in type from the original; put differently, sentence 125 does not say the same thing that 127 does. What is genuinely important about 125 is its paraphrasability as 128 (or 129) and the fact that the construction exhibited by 125 is restricted to certain kinds of nouns. Note the ungrammaticality of 130.

(125) I have a missing tooth.
(126) I have a tooth.
(127) I have a tooth and it is missing.
(128) My tooth is missing.
(129) One of my teeth is missing.
(130) *I have a missing five-dollar bill.

5.1.6. Note that in sentences 124 and 127, three things are involved: (*a*) a possessor (an 'interested person', to use the traditional term), (*b*) a body part, and (*c*) an attribute–(*a*) *me*, (*b*) *tooth*, and (*c*) *missing* respectively–and that the sentences provide alternate ways of ascribing the attribute to the possessor's body part. They are two distinct superficial ways of expressing the same relationship among these three concepts.

Using P, B, and A for *a*, *b*, and *c* above, we may represent the expression as seen in 125 as 131, and that as seen in 128 as 132.

(131) P^{nom} *have* [A → B^{acc}]
(132) [P^{gen} → B] *be* A

The same element, in other words, which in some of the paraphrases mentioned above appeared in either the dative or the genitive case forms now appears as the subject of the verb *have.* Bally, in fact, speaks of the invention of the word *have* as fulfilling precisely the function of allowing the *personne intéressée*, which otherwise would have to appear either in dative or genitive form, to become the subject of a sentence. Examples of all three surface appearances of a first person possessor are given by Bally (1926, p. 75) as 133–135. sentences 134 and 135 correspond to expression types 131 and 132 respectively; the expression type exemplified by 133 is given as 136.

(133) *Mihi sunt capilli nigri.*
(134) *J'ai les cheveux noirs.*
(135) *Mes cheveux sont noirs.*
(136) P^{dat} [B^{nom} *be* A]

5.1.7. Henri Frei (1939) surveyed this variety of surface representations of the 'same' sentences and added a fourth type, a type intermediate, in a sense, between that suggested by Formula 136 and that of 131. His example was sentence 137 (it also provided the title of his paper), which exemplifies the expression type we may wish to represent as 138.

(137) *Sylvie est jolie des yeux.*
(138) P^{nom} *be* [A $B^{oblique}$]

Frei points out that the construction seen in 137 is related to the category of inalienable possession, since while 139 and 140 are acceptable sentences, 141 and 142 are not.

(139) *Elle est fine de doigts.*
(140) *Elle est bien faite des jambes.*

(141) **Elle est fine d'ètoffe.*

(142) **Elle est bien faite des vêtements.*[61]

5.1.8. Since Frei sees this diversity as resulting from the attempt to 'condense' two judgments into one sentence–the two judgments that *P has B* and that *B is A* (in our terms)–he relates the constructions in question to the much discussed 'double subject' constructions of Japanese. In one type of this latter construction, two nouns appear before a verb or adjective, the first followed by the particle *wa* (indicating what I have called 'secondary topicalization'), the second by the particle *ga* (the particle of 'primary topicalization'). (Variations in the order and in the choice of particles do not change the status of the construction; the form described is the one most stylistically neutral.) The second of these nouns is of the inalienable type; the first identifies the object with respect to which the object identified by the second noun is 'inalienable'. The hackneyed example of the double subject construction is 143, a sentence which has 144 as a sort of forced paraphrase. In 144, the particle *no* is the particle whose functions are closest to those we would be inclined to label 'genitive'.

(143) *Zoo wa hana ga nagai.* 'Elephant *wa* nose *ga* long.'

(144) *Zoo no hana ga nagai.*

5.1.9. That expressions involving entities viewed as being closely associated with an 'interested person' have unique grammatical properties has also been observed in certain semantically unmotivated uses of 'reflexive pronouns' and the parallels one finds between these and various uses of the 'middle voice'. The connection with dative forms is seen in the fact that in some languages a kind of 'dative reflexive' is used in these special situations. Note 145 and 146.

(145) *Se laver les mains.*

(146) *Ich wasche mir die Hände.*

---

61. Frei (1939, p. 188). The expressions are limited to clear relational nouns, not only to body parts. Frei notes such phrases as 'des couloirs spacieux et *bas de plafond*' and 'libre de moeurs'. He beautifully demonstrates the distinctness of the sentences involving inalienable possession from overtly similar sentences of different grammatical structures with the contrast between i and ii below (p. 186).

(i) *La salle est pleine de visages.*

(ii) *La femme est pleine de visage.*

The connection between this use of the 'reflexive' and the category of inalienable possession is indicated by Bally, who points out that in item 147, *jambe* is the inalienable entity, while in 148 the word *jambe* can only (or, depending on my informants, can also) be understood as some independently possessed object, such as the leg of a table.

(147) *Je me suis cassé la jambe.*
(148) *J'ai cassé ma jambe.*

Notice that the *jambe* which does *not* have the possessive adjective is the one which is grammatically characterized as 'obligatorily possessed' (Bally, 1936, p. 68)!

5.2. *Adnominal Datives*

One way of introducing a possessive modifier of a noun has already been suggested: a sentence which could on its own assume the form '*X has Y*' is embedded to NP. Since it is desirable for an embedded sentence to have a semantic interpretation that contributes to the meaning of the whole sentence, the sentence-embedding source of possessives is needed as an explanation for alienable possession. In other words, one is satisfied to have the meaning of 149 represented as a part of the meaning of 150, though we may reject such a relationship between 151 and 152.

(149) I have a dog.
(150) my dog
(151) I have a head.
(152) my head

A distinct method is required for introducing the possessive element in the case of inalienable possession, a method which reflects the fact that the relationship between the two nouns in 'inalienable possession' is not (*pace* Frei) a sentential relationship.

For the types of inalienable possession that we have considered so far–in which the relationship has always been to an animate or 'personal' entity–the solution is to say that some nouns obligatorily take D complements. This can be managed by adding to the grammar another way of writing NP, namely the rule in 153.

(153) NP → N (D)

In the way that frame features for Vs relate to environments of Vs provided by the constituent P, frame features for Ns relate to environments provided by the constituent NP. It was suggested above that Ns which obligatorily take S complements are assigned the feature +[____S]. We may now add that Ns which obligatorily take D complements are characterized as having the feature +[____D]; and these are the inalienably possessed nouns. The notation imposes a subclassification of nouns into those which require adnominal D (such as *son*, *child* in the meaning 'offspring', German *Mann* in the meaning 'husband') and those which reject adnominal D (such as *person*, *child* in the meaning 'very young person', *Mann* in the meaning 'man').

The two sources of possessive modifiers which the grammar now makes possible (adnominal D and adnominal S of a certain type) provide the deep-structure differences needed for determining the difference in the form of the possessive modifiers in those languages which make the distinction overt in that way. Where further distinctions are made (as between body parts and kinship terms), the information on which such distinctions need to be based may be included as lexical features of the Ns themselves.

The general configuration of NPs containing Ds, then, is that shown in 154.

(154)

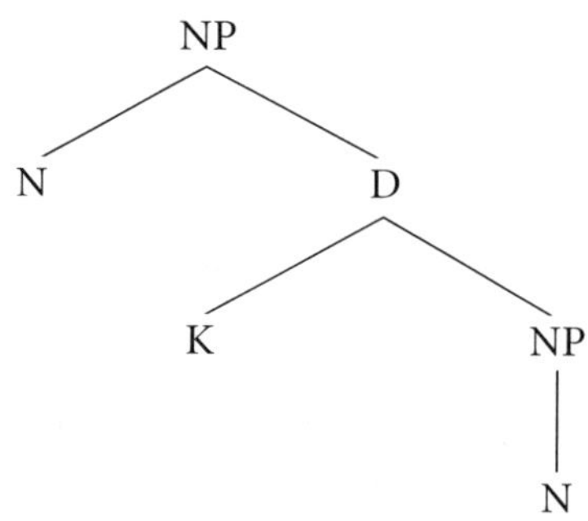

In some cases the adnominal D remains in the NP and in fact retains the surface features associated with D, as in 155; typically, however, a D inside an NP is changed to a genitive form, as in 156.

(155) secretary to the president
(156) the president's secretary

If determiners are universal,[62] then the expansion of NP must make provisions for them; but if they are not, then languages which have them will need 'segmentalization' rules of the type described by Postal (1966). At any rate, the determiners (which I represent as 'd') will figure in the various things that can happen to adnominal D. Sometimes, for example when a D remains in the NP without undergoing genitive modification certain of its features are copied onto the determiner so that the determiner may eventually assume the form of the appropriate 'possessive adjective'. This seems to account for such expressions as the possessive dative with kinship terms seen in some German dialects (recall 121) in Ossetic (see Abaev 1964, p. 18).

5.3. *Some Illustrations*

The D constituent often need not remain in the NP: under some conditions it may be 'promoted', so to speak, from the status of a modifier of an N (which it is in the deep structure) to the status of a major constituent on the next higher level of the syntactic structure. This can be seen in sentences having the base configuration [V + L + A]: just in case the N under L is a body part, the D which in the deep structure is subjoined to L is 'promoted' to become a constituent of P, yielding a sentence superficially of the type [V + D + L + A].

The verb *pinch* is accepted into the case frame [____L + A], and except when it has taken on the feature [+ passive], it is a verb which deletes the preposition of the following constituent. Let us consider sentences derivable from the deep structure seen in 157.

62. I am inclined to think that they are. See Fillmore (1967).

(157)

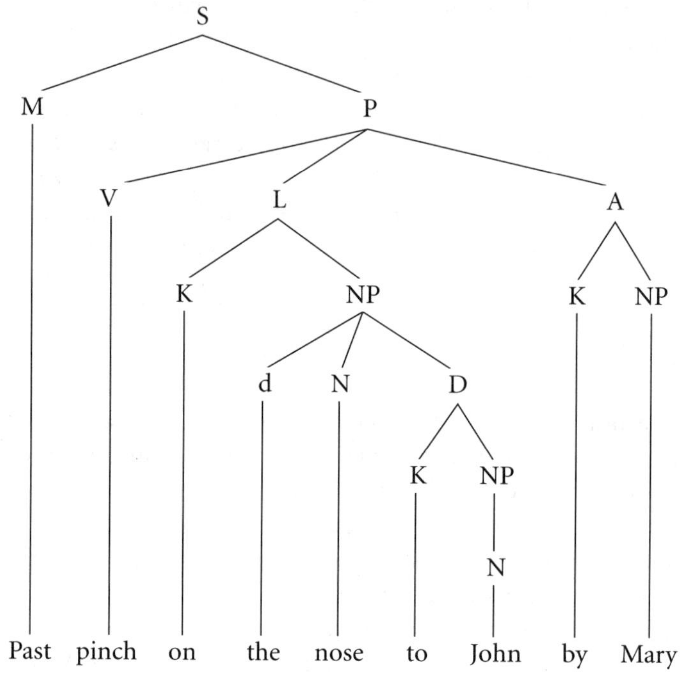

We shall see what happens to the sentence under four conditions: when the D remains inside L and A becomes the subject; when the D remains inside L and L becomes the subject; when the D is promoted and A becomes the subject; and when the D is promoted and D becomes the subject.

Whenever D remains inside NP (in this sentence), it is preposed to the N and converted to its genitive form, displacing the original determiner. Since it is a personal noun, the K element assumes the form of a genitive suffix. With nonpromoted D, in other words, 157 eventually becomes 158.

(158)

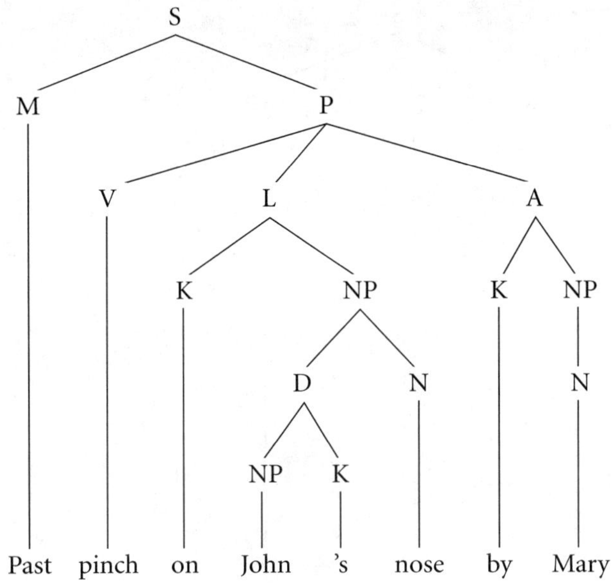

Diagrams 159–162 show the development from 158 when A is made the subject: the subject preposition is deleted and its case category is erased; the preposition after *pinch* is deleted and the case category L is erased; and the tense is absorbed into the V.

(159)

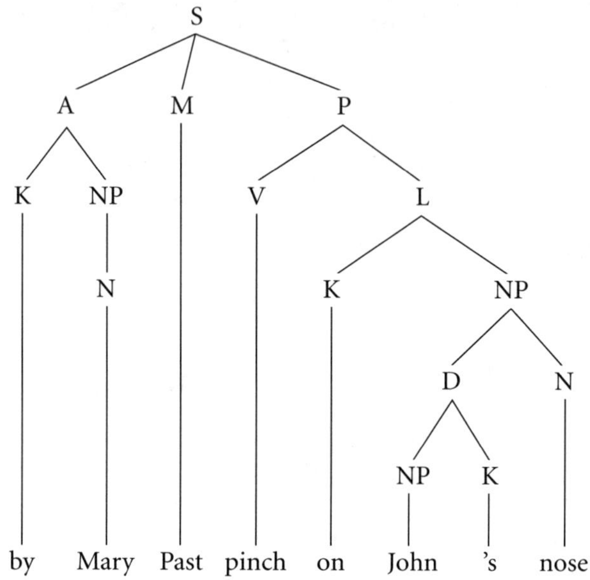
S
A M P
K NP V L
N K NP
D N
NP K
by Mary Past pinch on John ’s nose

(160)

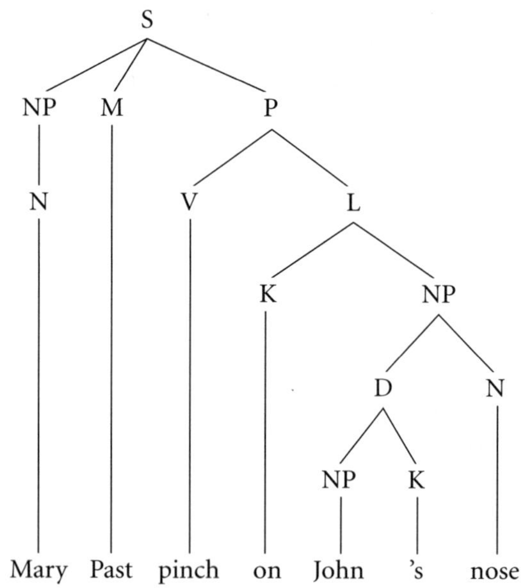
S
NP M P
N V L
K NP
D N
NP K
Mary Past pinch on John ’s nose

(161)

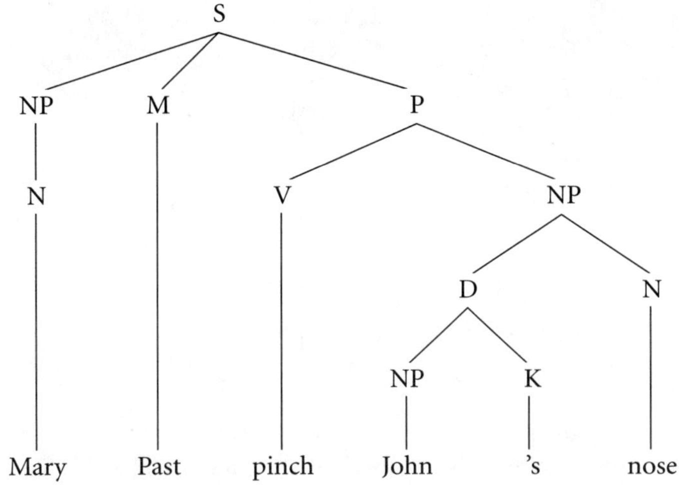

(162)

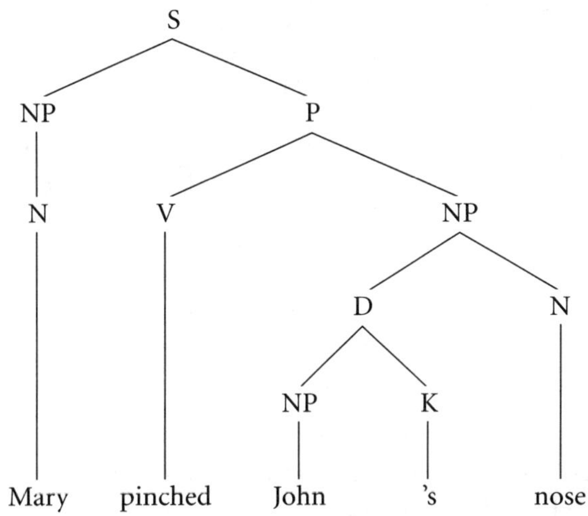

If the L of 158 is chosen as subject instead of the A, the result is 163. This choice of subject requires the V to assume the feature [+ passive], which causes it to lose its ability to delete following prepositions and its ability to take tense affixes. The surface structure eventually resulting from 163 is 164.

(163)

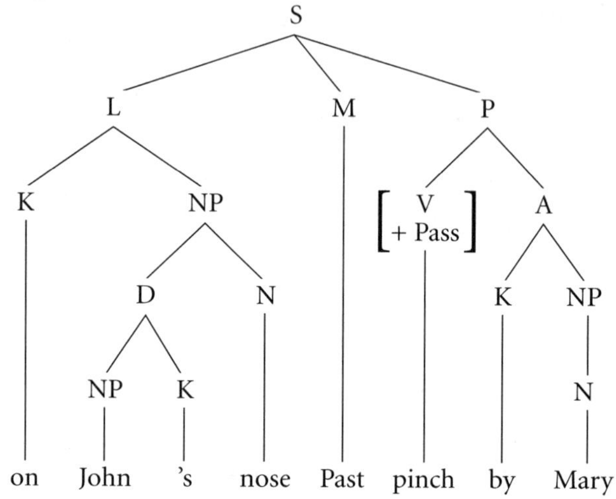

(164)

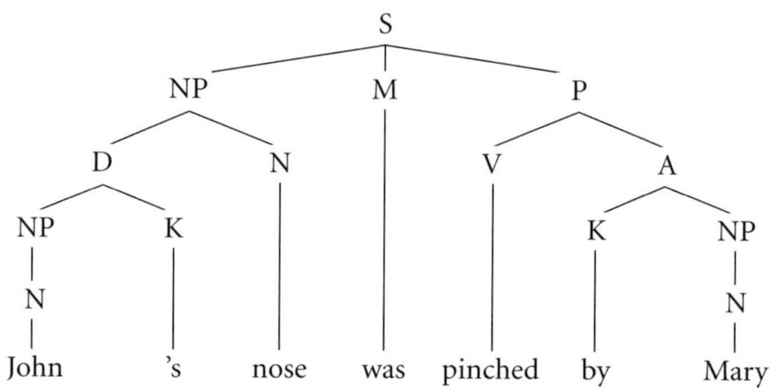

Backing up to 157, we may now see the consequences of ‘promoting’ adnominal D. When the D is removed from L and becomes the left-most case constituent in P, the resulting structure is 165.

(165)

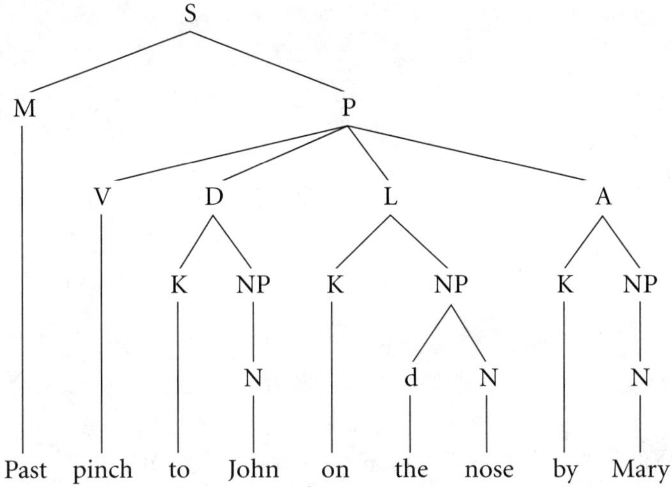

The possible subjects for 165 are the A or the newly promoted D. When the subject is A, we get 166, a structure which, on application of the rules we have learned, eventually becomes 167.

(166)

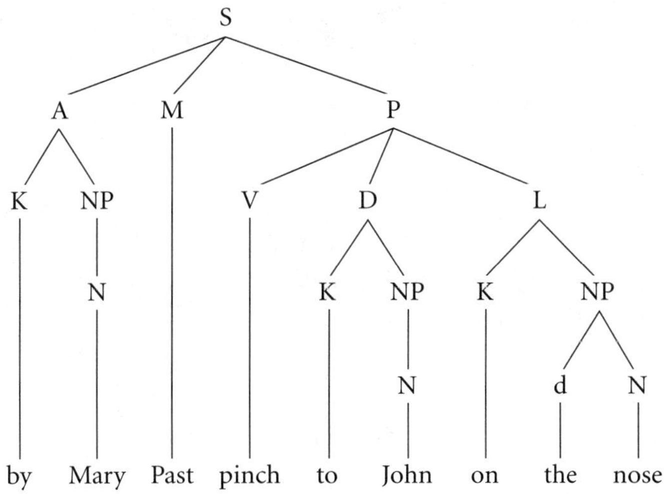

(167)

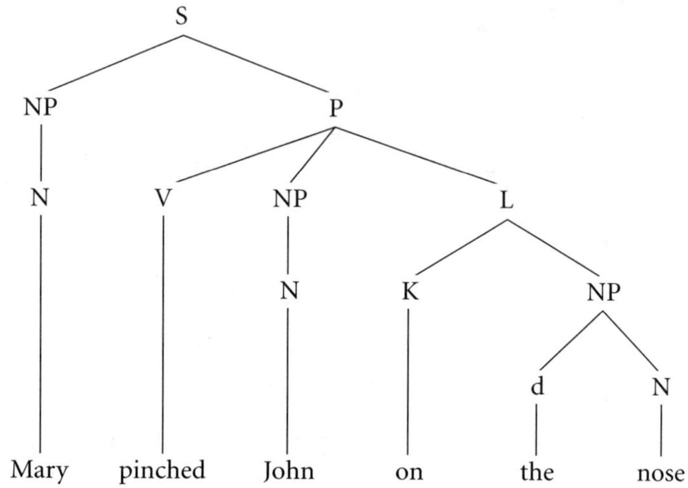

When D is made subject, on the other hand, we get 168; on applying the rules appropriate to a V with the feature [+ passive], we eventually get 169.

(168)

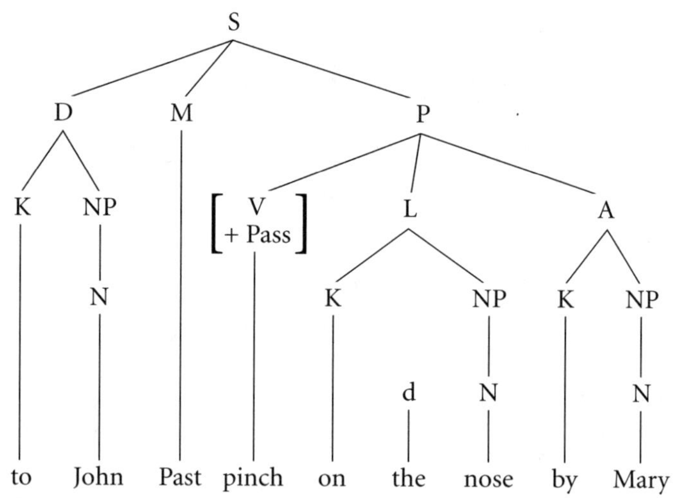

(169)

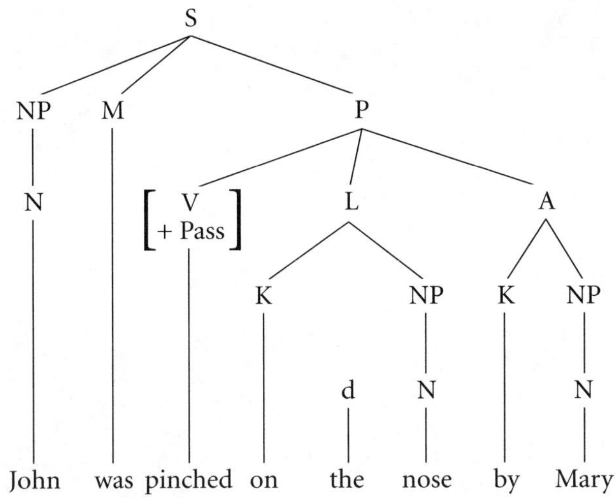

We may turn to the problem which interested Bally and Frei and examine the role of adnominal D in sentences which assign attributes to obligatorily possessed elements. The basic structure of such sentences can be illustrated by 170.

(170)

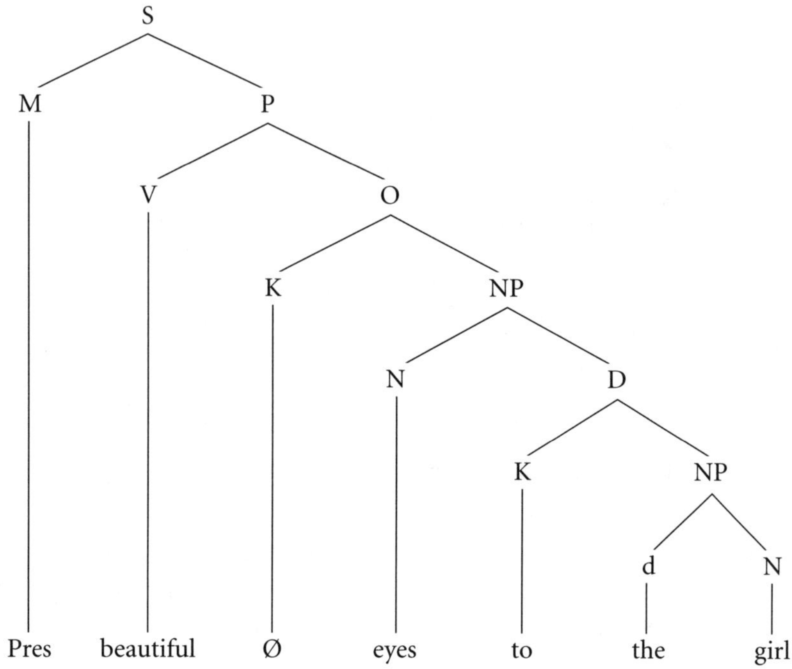

In languages which allow the D to remain in the NP, the D element is converted to its genitive form. In English this results in 171. Since 171 has only the form [V + O], the O is necessarily chosen as subject, and the result for English is 172.

(171)

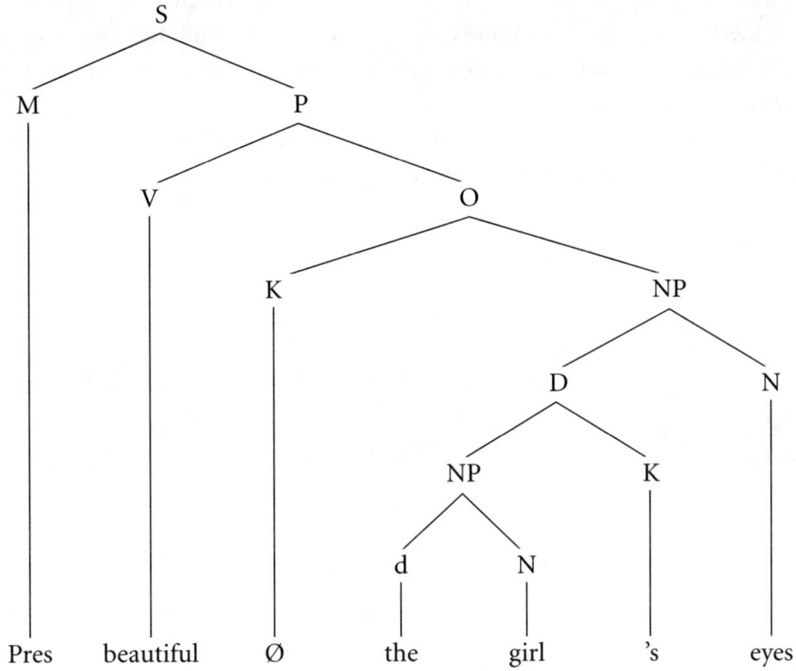
S
M
P
V
O
K
NP
D
N
NP
K
d
N
Pres
beautiful
Ø
the
girl
’s
eyes

(172)

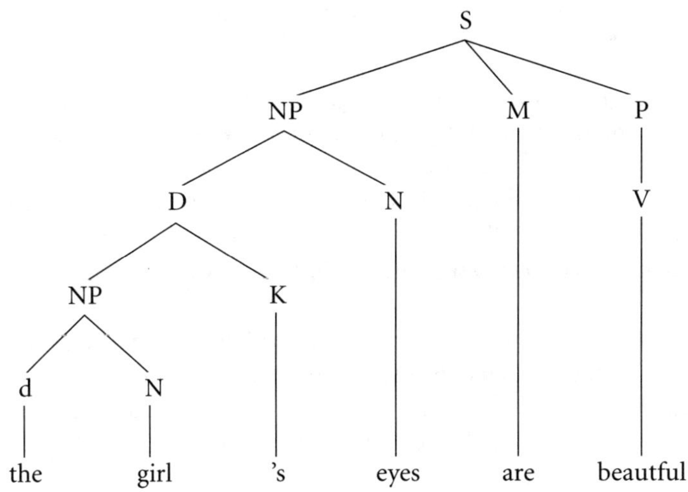
S
NP
M
P
D
N
V
NP
K
d
N
the
girl
’s
eyes
are
beautful

Notice that since the V is an adjective, it is incapable of 'absorbing' the tense,[63] so that it requires the provision of *be* within the M constituent. Example 172 is a rendering of 171 in which the V is predicated on the O and the D is subjoined to the O. Thus it is analogous to our earlier sentences 128 and 135 and is of the type indicated in 132.

Suppose next that the D of 171 does get 'promoted'. The result of introducing the D in this way as an immediate constituent of the P is 173.

(173)

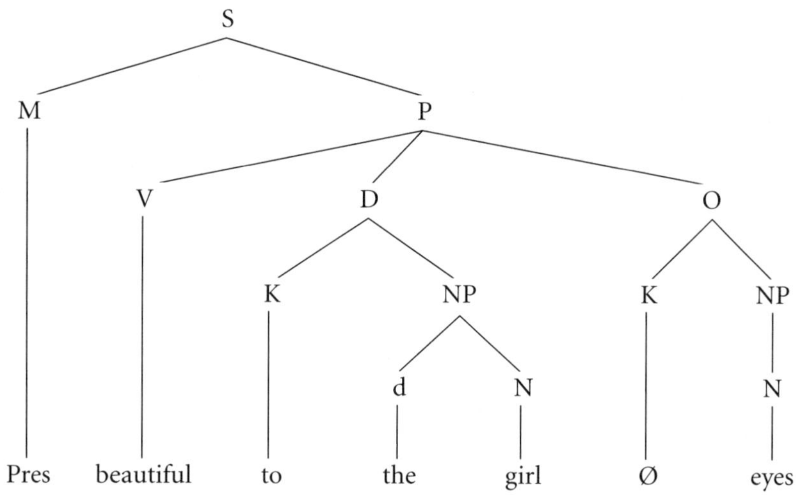

Some languages allow the O of Configuration 173 to become the subject and the D element to remain in the expected surface form for D, as in sentence 133. Others allow the D element to undergo secondary topicalization when the O is subject, resulting, for example, in one case of the 'double subject' construction in Japanese (recall 143). The general expression type for sentences resulting from 173 when O becomes subject is suggested by the formula in 136 above.

Many languages allow the D to become subject. When this happens and there are no other changes, the O appears in some oblique case form. This is so because, since *beautiful* is not a true verb, the body-part word cannot be converted into an 'object'. The initial structure is seen in 174; it is one which is

63. Stated more accurately, Vs which are adjectives, passives, or progressives are incapable of absorbing the right-most affix in M.

not typical of English, though it is perhaps seen in such expressions as those given in 175 and it may represent a stage in the derivation of a phrase of the type given in 176.

(174)

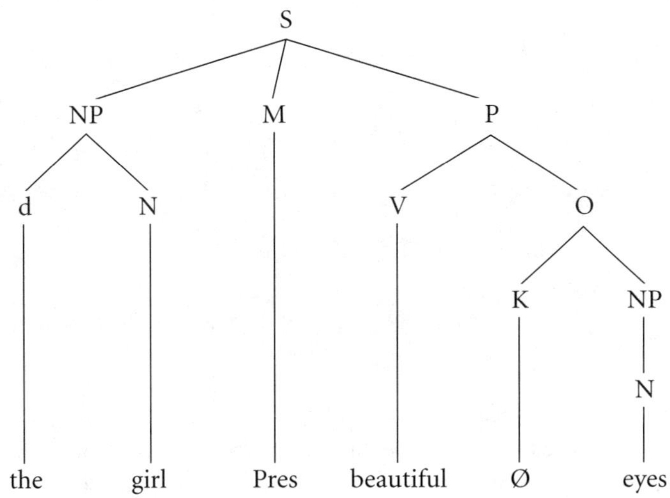

(175) tall of stature, blue in the face

(176) broad-chested, fat-legged

It appears to be the structural form underlying 137, whose expression type is given in 138. The construction is apparently quite rare in French; Frei speaks of it as a 'short-circuited' version of the sentences with *have*.

Another possibility, when D is subject, is to attach the adjective to the NP indicating the body part. I propose, in an unhappily quite ad hoc fashion, that this be done without removing the constituent label V. I believe there are some arguments for retaining at least an abstract V under P at all times. This constraint may turn out to be better motivated than it seems, for this structure appears to reflect what was needed in those languages which adopted a verb like *have*.

The structure I have in mind is that shown in 177.

(177)

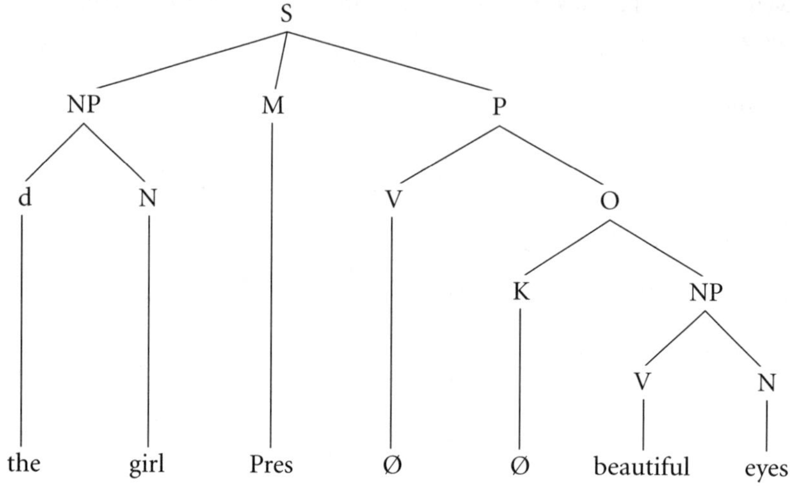

With the V under P vacant, a *be* must be added to the M in languages which allow structures of this type to become sentences directly. Notice that in this construction the modified body-part NP as a whole is 'in' some case form. The formula for this expression type has not yet been given; it would be something like 178.

(178) P *be* [A → B]^{oblique}

Conceivably this is the structure underlying such predicates as those shown in 179; the difference between predicates of the type in 138 and those of the type in 178 is seen in the Latin paraphrases of 180 and 181 respectively.

(179) of tall stature; *di bello aspetto*
(180) *aequus animo*
(181) *aequo animo*

The last possibility, then, is to insert into the vacated V position the function word *have*, a verb which takes the modified body-part noun as its 'object'. In English, we have seen, this involves deleting the preposition. The result of modifying 177 in this way is 182.

(182)

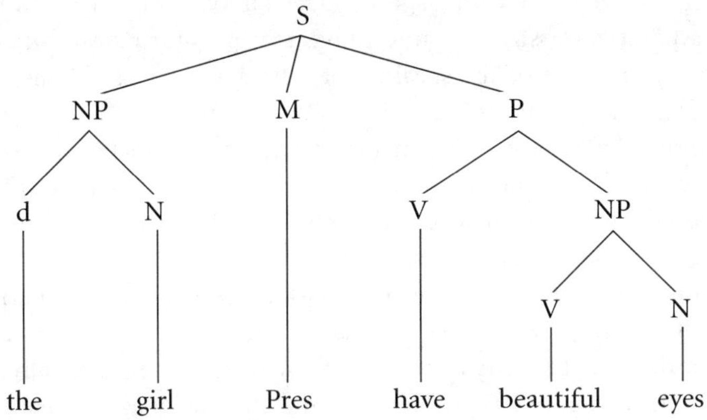

In short, it appears that the considerable surface variety found in sentences involving attribution of some property to an inalienable noun is to be accounted for by positing for universal grammar, in the spirit of Bach (1965), a set of recurrent transformations which each language uses somewhat in its own way. For sentences of the general structural type 183

(183) $^{P}[V^{O}[D + N]]$

where V is an adjective and N is a body-part noun, the options are *a* to *d* below:

(a) promote D
(b) choose D as subject
(c) copy adjective into body-part NP
(d) insert *have* into the vacated V

When *a* is not applied, the D becomes a genitive modifier of the body-part N and the whole O becomes the subject. When *b* is not applied, the O becomes the subject. When *c* is not applied, the 'short-circuited' sentences of Frei are the result. Rule *d* is available only to languages which have 'invented' *have.*

5.4. *Further Remarks on Inalienable Possession*

If the feature of inalienable possession is to be treated as a universal property of language, then either vocabulary items which are translations of each other will be categorized alike with respect to alienability, or the ways in which languages separately classify the 'same' things may possibly reflect differences in the psychic make-up of the speakers of different languages. Many scholars have seen in the data on inalienabilia an opportunity for the science of language to shed light on primitive mentality and on the possible range of man's concept of 'self'. Since the differences appear more and more to be differences on the level of surface structure, it may be advisable to wait some time before reaching any conclusions on these matters.[64]

Adnominal D will certainly be needed for more than body-part nouns and names of relatives. Directional indicators like *right* and *left* are probably nouns of this type too. The reason that these words appear typically without any personal reference in English and many other languages is that they frequently refer to position or direction with respect to the speaker or addressee of the utterance, and there are simply many situations in which an adnominal D does not need to be expressed if it identifies speaker or hearer.

There are, too, many relational nouns which do not have a specifically personal reference. We might wish to say that certain 'locational' nouns take an adnominal L. These nouns sometimes name parts of the associated objects, as in 184, and they sometimes identify a location or direction stated with reference to the associated object but not considered as a part of it, as seen in 185. 'Nouns' of the second type appear superficially as prepositions in English.

(184) corner of the table, edge of the cliff, top of the box
(185) behind the house, ahead of the car, next to the tower

6. Problems and Suggestions

There is a considerable residue of unsolved problems in the grammatical description of language phenomena, and it is disappointing though not surprising to realize how many of them remain unsolved under the formulation of grammar I have been suggesting. Those which come most quickly to mind are coordinate conjunction, nominal predicates, end 'cognate objects'.

64. For representative statements on the sociological relevance of the study of inalienable possession, see Lévy-Bruhl (1916, p. 103), Bally (1926, p. 68 *et passim*), Frei (1939, p. 192), and van Ginneken (1939, p. 90). For a catalogue of noun classifications based on grammatical differences associated with inalienable possession, see Rosén (1959, p. 268 f.).

6.1. *Coordinate Conjunction*

There may be a relationship between the ways in which languages deal with 'comitative' constructions and the phenomenon of coordinate conjunction of NPs. Put in case terms, there may be a relationship between conjunction of NPs and what one might wish to refer to as a comitative case. Jespersen (1924, p. 90) noticed the parallels between *with* (a preposition which has a comitative function) and the conjunctor *and*, as in such pairs of sentences as 186 and 187.

(186) He and his wife are coming.
(187) He is coming with his wife.

Japanese has separate devices for indicating sentence conjunction and NP conjunction, and the postposition used for NP conjunction is identical with the comitative postposition. In a conjunction of NPs, all but the last have the postposition *to*. The last one has the postposition appropriate for the case role of the whole NP. Compare 188 and 189.

(188) *Tanaka-san to Hashimoto-san ga kimashita.*
'Mr. Tanaka and Mr. Hashimoto came.'
(189) *Hashimoto-san ga Tanaka-san to hanashimashita.*
'Mr. Hashimoto spoke with Mr. Tanaka.'

Redden points out that in Walapai a sentence has only one noun in the 'nominative' case. Noun conjunctions are effected by having the 'ablative' suffix—the suffix with comitative function—on all but one of the nouns in a conjunction. Thus, in 190, /*-č*/ is nominative, /*-m*/ ablative.

(190) /hàtθáùač hmáɲm/
'the dog and the boy' (lit. 'the dog with the boy')

It may be that the rule in 191 is needed as an expansion rule for NP.

(191) NP → NP+ C

Using X as a cover term for the various case categories, 191 will produce such structures as 192.

(192)

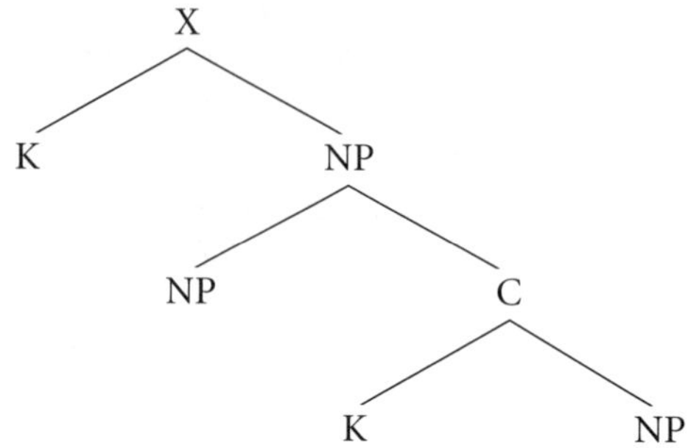

The case category C has a very special status, since the selectional constraints on nouns under C are those of the superordinate NP. What is needed, in other words, is a rule which imposes on any N under C the same redundant features that are associated with the dominating non-C case.

A subjoined C under some circumstances must remain in the large NP. In languages which lack a generalized conjunctor, the case marker is simply that appropriate to C (the postposition *to* in Japanese, the suffix *-m* in Walapai); in languages which have a generalized conjunctor, this word replaces the case marker, in the way that *and* replaces *with* under certain conditions.

The structure underlying 186 and 187, then, might be something like 193; we ignore the source of *his*.

(193)

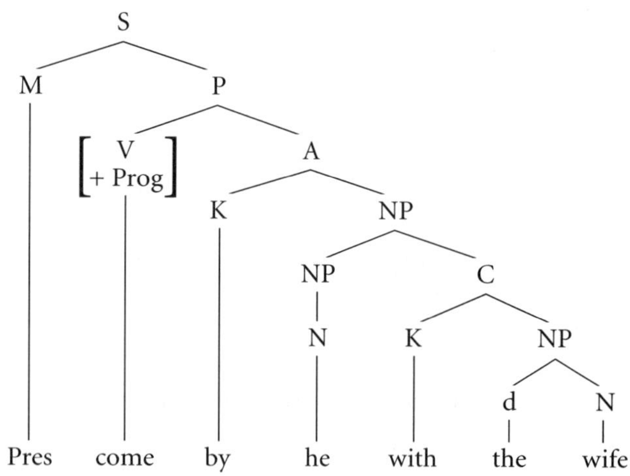

If the C remains inside the NP, the entire A becomes the subject, yielding sentence 186; if the C is promoted, however, as in the structure shown in 194, it is left behind when the A becomes the subject, resulting in sentence 187.

(194)

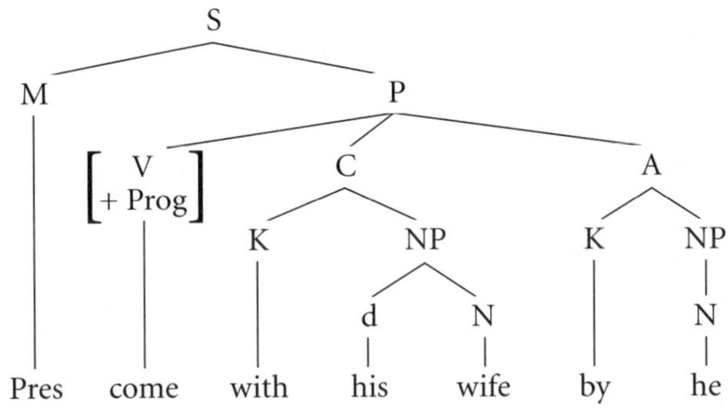

It is quite unlikely that the numerous problems associated with NP conjunction can be appreciably simplified through this approach, but that there is some connection between conjunction and comitative uses of NPs cannot be doubted. Lakoff and Peters (1966) have recently presented very persuasive arguments that the 'direction' of the relationship is the opposite of what I have suggested; that, in other words, comitative phrases are derived from NP conjunction rather than the other way around.

6.2. *Nominal Predicates*

Nothing that has been said so far suggests a way of providing for sentences of the *N be N* type. It is clear that they represent a distinct sentence type from those involving any of the case relations discussed above, though more than one case relationship may be involved in these sentences. (The terms *essive* and *translative* come to mind.)

Some nouns that appear in predicate position are restricted in their occurrence elsewhere. It might be possible to treat these nouns as, on one level, Vs which are restricted to the form [____A]. Examples are words like *idiot*, *bastard*, and *fool*. The environment contains A because the subject is always animate and because the constructions exhibit selectional and transformational properties associated with Vs having As in their environment. Notice 195 and 196.

(195) Don't be a fool.
(196) He's being a bastard again.

This interpretation appears to account for the fact that we have sentences like 197, but not–with *idiot* used in this same 'evaluative' sense–198.

(197) John is an idiot.
(198) An idiot hit the first homerun.

Further evidence that the word is properly treated as a V is found in the fact that these nouns may accept types of modification usually associated with adjectives, as in 199.

(199) John is quite an idiot.

The serious problems are (*a*) with the use of words like *idiot*, *fool*, and so forth, in other contexts, as in 200, and (*b*) with the use of non-evaluative Ns in predicate sentences, as in 201.

(200) That rat swiped my lunch.
(201) That boy is my nephew.

A new case category or two could be invented for the occasion, of course, but such matters as the requirement that subject and predicate NPs agree in number remain as serious as they ever were. Perhaps some solution is forthcoming along the lines of Bach's proposals in Bach (1968).

6.3. *Cognate Objects*

A difficulty of another sort is presented by the so-called 'cognate-object' constructions. These are constructions in which, at the very least, there is a high selectivity between a specific V and an 'object' N, and in which the V + N combination in one language might well be matched by a V alone in another.

Slightly modifying a recent analysis by Sandra Babcock (1966),[65] I would propose that there are contexts in which the case category F (factitive) may be left lexically empty, and that certain words classified as Vs may be inserted

65. Compare too the interpretation in terms of 'quasi transformations' found in Harris (1957, Section 30).

specifically into frames containing dummy Fs. These words may have associated with them special N representatives (for example, *bath*) and special pro-Vs (for example, *take*). The rules that apply to dummy-F sentences are the following:

(a) Copy the N-representative of the V under the F.
(b) Replace the V by the designated pro-V.

The rules may have separate conditions of optionality for different Vs. The cognate-object V *dream* may appear as a V in its own right, or it may appear in dummy-F sentences. As a cognate-object verb, it has *dream* as its N-representative and *have* as its pro-V; it is further specified as selecting either the preposition *about* or *of* for the O constituent and as not requiring Rule *b*.

When the N-representative associated with *dream* is copied into the F constituent, the result is sentence 202; when the associated verb *have* replaces the V, the result is 203.

(202) John dreamed a dream about Mary.
(203) John had a dream about Mary.

With these devices, we may in fact consider extending the interpretation of cognate-object constructions in the following way. Some words may be treated as cognate-object Vs even though the rule for replacing the pro-V is obligatory. The V *nightmare*, for example, might have *nightmare* listed as its N-representative and *have* as its pro-V. Thus, on applying Rule *a*, the structure in 204 becomes the intermediate structure in 205; on applying Rule *b*, 205 is converted to 206. Analogous uses of this device could possibly account for the connection between *suggest* and *make a suggestion*, *shove someone* and *give someone a shove*, and so on, but many serious problems remain. In particular it is not obvious how sentences like 207 and 208 can be dealt with in accordance with these proposals.

(204)

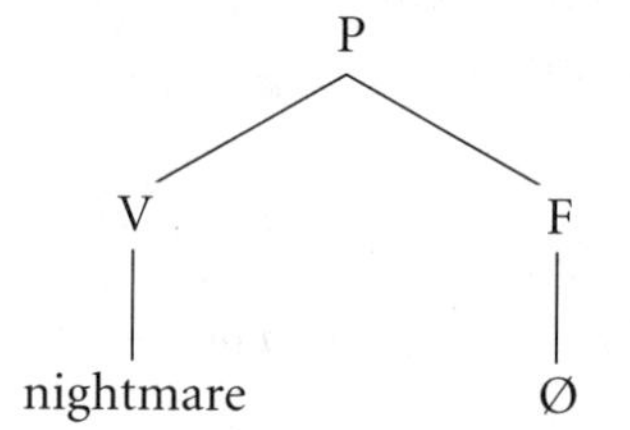

(205)

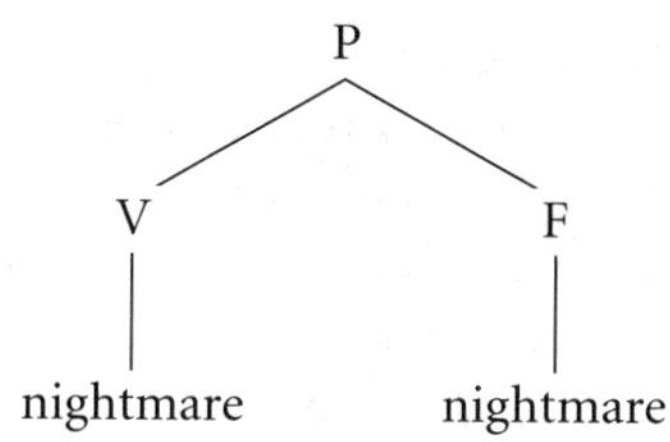

(206)

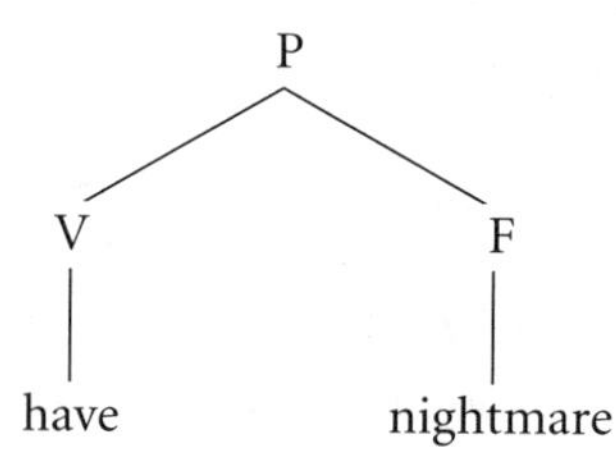

(207) She made several ridiculous suggestions.

(208) I had a terrible nightmare last night.

6.4. *Other Problems*

There are many issues for which I cannot even pretend to see solutions. The apparent connection between surface cases and 'partitive' functions; the restriction of 'definiteness' in some languages to NPs in particular surface-case relations (typically, the 'direct object'); the extreme variety of surface realizations of the same meaning (from the same deep structure?) that Jespersen illustrates in connection with what he calls 'rank shifting' (1924, p. 91), to name just a few.

The difficulties mentioned so far are empirical in nature, but many formal problems exist as well. One of these is whether the permitted arrays of cases under P need to be generated via phrase-structure rules, since one of the most important functions of the PS rules has been that of defining grammatical relations–that is, that of defining phenomena which are here partly treated categorially rather than configurationally. Related to this problem are the apparent dependency relations that exist *among cases*. It appears, for example, that the occurrence of B (benefactive) phrases in a sentence has more to do with whether the sentence contains an A than with independent specific properties of Vs. One is almost willing to allow these facts to be expressed by a generative process which chooses a verb, then the cases required by that verb, then the other cases that are compatible with the cases originally chosen. The issue is not whether the permitted sequences can or cannot be generated by PS rules–there is no doubt that they can–but whether the kinds of co-occurrence or dependency relationships that seem to obtain might not be more efficiently stated in some other way. (Modifications of transformational grammar of the type introduced in Chomsky (1965) made it no longer necessary to use PS rules for subclassification of lexical categories or for the choice of lexical items. If the provision of syntactic relations of certain kinds must also be handled by some device other than PS rules, there is a chance that rules of this type may be abandoned altogether.)

Whether the cases should be represented as categories dominating NPs or in some other way is an issue which seems to me to be fairly wide open. One advantage of the categorial treatment is that NPs made subject and object may be said to have lost their 'original' case relation to the sentence (by the rule which 'erases' the case category whenever the case marker K has been deleted–that is, a 'node-razing' rule) with the result that their form can only be determined by referring to their 'pure relational' status. Thus it would appear that the surface distinction between labeled and configurationally defined relations on NPs may correspond to the traditional distinction between the 'concrete' and the 'grammatical' cases. (How the genitive figures in this distinction is not clear under either interpretation.)

Several people have pointed out to me the apparent convertibility of underlying representations in case grammar into objects which resemble dependency diagrams and tagmemic formulas. If the K elements are interpreted as constituents of NPs, then the case categories *unarily* dominate NPs. This makes them equivalent to *labels on the branches that link P with the various NPs that are directly related to it.* If the only function of the P is to provide a constituent in terms of which the NPs can be related to the V, one may just as well represent these relationships more directly by replacing the node P by the V. The result is no longer a diagram of constituent structure, since lexical elements are inserted into 'dominating' nodes; but it may turn out to be just as possible to represent the needed constituent organization of sentences from a 'stemmatic' diagram of the type used by Tesnière or Hays, as from a phrase-structure tree diagram.

There is an easy conversion from underlying representations of case grammar to 'tagmemic' formulas, too, as long as the case categories *unarily* dominate NPs. Or, for that matter, a case-grammar diagram could simply be read off as a tagmemic formula, as long as certain symbols were designated as function indicators. One can as easily say 'NP filling an A slot' as anything else. The crucial difference between the modification of transformational grammar that I have been suggesting and the typical tagmemic study is in the insistence here on discovering the 'deepest' level of the 'deep structure'.

7. Closing Words

One criticism of case grammar that has been brought to my attention is that it is too strongly motivated by semantic considerations. Many of the analyses have (hopefully) the result that certain semantic distinctions and interlanguage commonalities are revealed in fairly direct ways in the deep structures of case grammar, but, it has been argued, syntactic analyses should be based on syntactic data alone and on one language at a time.

The question arises whether there is a 'level' of syntactic description that is discoverable one language at a time on the basis of purely syntactic criteria. If it is possible to discover a semantically justified universal syntactic theory along the lines I have been suggesting, if it is possible by rules (beginning, perhaps, with those which assign sequential order to the underlying order-free representations) to make these 'semantic deep structures' into the surface forms of sentences, then it is likely that the syntactic deep structure of the type that has been made familiar from the work of Chomsky and his students is going to go the way of the phoneme. It is an artificial intermediate level between the empirically discoverable 'semantic deep structure' and the observationally accessible surface structure, a level the properties of which have more to do with the methodological commitments of grammarians than with the nature of human languages.

3

The Grammar of Hitting and Breaking

This paper picked up some observations of case frame *differences that had been introduced in "The Case for Case", but presented them as part of a plan of argumentation about lexical meanings and the relation between lexical meaning and the syntactic structure of sentences containing words with those meanings. A key distinction is one between verbs of change of state (and caused change-of-state) and verbs of impact.*

Among the ideas emphasized in this paper are these:

(1) The core sense of a verb can be independent of the valence possibilities of the verb, since the presence or absence of semantic cases can themselves contribute to the meaning. Thus in the three "uses" of the verb break, *as in* I broke the window (with a rock), the rock broke the window, *and* the window broke, *each entails the same consequence for the affected window, and they differ only in ways signaled by the presence or absence of mention of an agent or an instrument.*

(2) There is a difference between the roles NPs can play in their sentence and the "entities" that participate in the eventuality designated by the sentence. First there is a distinction between the number of roles and the number of entities: for example, in contrasting the boy broke the window *and* the boys broke the window, *there is a single agentive role in each sentence but more than one agent in the second sentence. Second, there is a distinction between the roles played by participating entities and their* sorts: *the same kind of thing can participate in more than one role, as in the interchanged roles of* Jimmy hit the rock *and* the rock hit Jimmy. *A lot of mail I received in the early seventies suggested that what I had written confused these notions.*

(3) A distinction is made here between the kind of semantic observations that a linguist can make, where the distinctions need to have clear syntactic consequences of some sort, and distinctions that should be left to the lexicographer, because they can be ultimately unclear or can vary from user to user. There are

Originally published in 1970

clear syntactically describable similarities that hold twist, spin *and* rotate *together, or* hit, smite, *and* slap, *but it should be left to lexicographers to worry about how these words are distinguished within each group.*

The contrast between implicit and explicit knowledge is perhaps nowhere greater than in man's use of language. If a person wishes to record and systematize what it is that he "knows" in knowing his language, he cannot simply write it down–he has to discover it first. And in attempting to discover the nature of his linguistic knowledge, he will find that satisfactory statements do not come easily. It is quite certain that the average adult speaker of English, even if given a year's time, simply could not come up with anything like a reliable explanation of how the word *ever* is used and what it means. And yet probably never in his adult life has this same person "made a mistake" in his use of this word, nor has he used it in ways inappropriate to his intentions.[1]

What a speaker of language knows about the individual "words" of his language and the conditions that determine their appropriate use is perhaps the most accessible aspect of linguistic knowledge, but at times it too is extremely subtle and–at least on the face of it–extremely complex. In this paper I hope to explore some of this subtlety and complexity by considering how speakers of English use and understand two very ordinary verbs, *hit* and *break*.

My treatment of these words, and the kinds of evidence I shall appeal to in uncovering their grammatical and semantic properties, will bear some similarity to the style of argument one finds in the writings of the so-called "ordinary language" philosophers, but with two important differences. The first of these is that, unlike words like *know, good, ought, real,* or *exist,* the words *hit* and *break,* in themselves, have no philosophical interest. The second is that a linguist's analysis of words cannot be accounted satisfactory until his observations can be incorporated into a general empirical theory of linguistic structure. In other words, we cannot be satisfied that our inquiry has been completed until we are convinced that the concepts and principles we have used in organizing our observations are proper to some substantively and formally specific explanatory theory of the nature of human language. What this means in practice is that the linguist keeps the connection with

1. This paper was written while the author was taking part in a research project supported by the National Science Foundation through Grant GN534 from the office of Science Information Service to the Computer and Information Science Research Center, The Ohio State University.

The author wishes to express his gratitude to Gaberell Drachman and David L. Stampe for many helpful suggestions.

empirical linguistic theory in mind, at least, whereas the philosopher traditionally has not been expected to do so.

A grammatical description of a language is successful if it accounts for precisely the facility that an ideal speaker of a language has in producing and understanding the grammatical sentences in his language. The knowledge that the speaker brings to bear in exercising this ability may be separated into the "general" and the "specific." One's "general" knowledge about a language is organized and displayed in its "grammar"; one's "specific" knowledge about the individual linguistic objects known as words or "lexemes" is collected and itemized in a "dictionary" or "lexicon" of the language. In this paper I shall attempt to determine at least some of the specific things that speakers of English know which account for their ability to use the words *hit* and *break* correctly. Put differently, the goal of this paper is to discover the information that needs to be registered, in one way or another, in the entries for these two words in a scientifically sound dictionary. It is in this sense an exercise in lexicological research.

The reader might at first be inclined to think that the task we have set for ourselves has already been completed–that anyone who wants to learn the lexical facts about *hit* and *break* can do so quite readily by looking the words up in a standard dictionary. It can be shown very easily, however, that there are indeed important facts about words which the makers of dictionaries do not generally bother to tell us. For example, if you look up the words *sick* and *ill* in a standard dictionary, you will be told that they are synonymous in one of their meanings, but what you will not be told is that although both of these adjectives can occur as predicates, only *sick* can occur attributively. Notice examples (1) and (2):

(1) The children are $\left\{\begin{matrix}\text{sick}\\ \text{ill}\end{matrix}\right\}$

(2) the $\left\{\begin{matrix}\text{sick}\\ \text{*ill}\end{matrix}\right\}$ children

Or if you look up the word *good*, you will not find out from the dictionary something that every speaker of English knows, and that every foreign speaker of English needs to know, that (apparently) *good* is the only adjective in English which can take, in negative and interrogative predicate sentences, the "quantifier" *any*. Notice examples (3) and (4):

(3) Is it any $\left\{\begin{matrix} \text{good} \\ {}^{*}\text{pink} \end{matrix}\right\}$?

(4) They weren't any $\left\{\begin{matrix} \text{good} \\ {}^{*}\text{tall} \end{matrix}\right\}$.

An ordinary dictionary will not tell us everything there is to know about *hit* and *break*; there is much we shall have to figure out for ourselves. One of the things it will tell us, however, is that each of these words has several senses. We shall concern ourselves here only with what might be called their "basic" or "nontransferred" meanings. We shall consider their use in expressions about *hitting trees* and *breaking sticks*, but we shall ignore their use in expressions about *hitting upon good ideas* or *breaking in a new man*, for example.

The first problem to take note of is that in a grammar which requires "subjects" in the deep structure[2] representation of every sentence, it is necessary to recognize three distinct verbs having the form *break* and two distinct verbs having the form *hit*, and that these distinctions are unaffected by the decision to restrict our attention to the basic meanings of these words.[3]

The three *break* verbs may be referred to as *break-1*, *break-2*, and *break-3*. The verb *break-1*, which is illustrated in sentence (5),

(5) The stick broke.

is an intransitive verb which asserts of its "subject" the particular deformation or change of state we associate with the meaning of the verb.

The verb *break-2* is seen in sentence (6):

(6) John broke the stick (with a rock).

It is used for asserting of an object the same change of state mentioned in connection with *break-1*, but *break-2* asserts it of its direct object. In general, precisely those noun phrases which can occur as subjects of *break-1* can appear as objects of *break-2*; *break-2* assigns an "agentive" or "instigative" role to its subject, which is typically animate. As shown in the parenthesized

2. For a statement of the distinction between "deep" and "surface" representations of the structure of sentences, see Chomsky (1966, pp. 31–51) and Chomsky (1965).

3. Although it is certainly possible to rephrase the facts under examination in such a way that there is *one* verb *break* with several distinct "uses." it is important to realize that the formal complexity that concerns us here is in no way affected by this reformulation.

expansion of (6), *break-2* may co-occur with a phrase which identifies the "instrument," i.e., the inanimate object immediately responsible for the action of breaking.

The verb *break-3*, seen in sentence (7),

(7) A rock broke the stick.

differs from *break-2* in accepting inanimate subjects[4] and in not permitting an instrument phrase in the same clause. In other words, a sentence like

(8) *A rock broke the stick with a hammer.

is unacceptable (where *A rock* is not intended metaphorically), because this verb interprets the role of its "subject" instrumentally, and a simplex clause presumably can identify only one "instrument."[5] Both *break-2* and *break-3* agree in the semantic relation to the direct object.

There are three break verbs, but there are only two verbs with the form hit (in the meaning we have in mind); and these are *hit-1*, seen in (9),

(9) John hit the tree (with a rock).

and *hit-2*, seen in sentence (10),

(10) A rock hit the tree.

Verb *hit-2* parallels *break-3* in assigning an instrumental role to its (inanimate) subject, and in not tolerating an instrumental *with*-phrase. That is, the sentence:

(11) *A rock hit the tree with a stick.

---

4. There are good reasons for saying that an instrumental noun phrase (in the intended sense) is always underlyingly inanimate. A sentence like *The dog broke the stick*, when it is used to refer to what happened to the stick when we threw the dog at it, must then be interpreted as containing in its deep structure the noun phrase *the dog's body*. The word *body*, then, being inanimate, is capable of fulfilling an instrumental function in the clause. There are languages in which the distinction between *the dog* as agent and experiencer and *the dog* (= *the dog's body*) as physical object must be made overt. Mohawk, Paul Postal has informed me, is one such language.

5. At most, a simplex clause identifies one "noun phrase" as having an instrumental role. This is not to deny that the noun phrase can be compound.

is unacceptable. And *hit-1* parallels *break-2* in assigning an agentive role to its (animate) subject and in accepting an instrumental phrase [as in the parenthesized expansion of (9)]. The two verbs *hit* agree in their semantic relation to the direct object.

Our two sets of verbs differ in that there is no intransitive verb *hit* corresponding with *break-1*, since there is no sentence of the form (12).

(12) *The tree hit.

This division of *break* into three verbs and *hit* into two verbs is necessary if we wish to include in our descriptions of what we know about individual verbs constraints on the noun phrases that can occur in construction with them, the sentence types in which they can play a role, and the semantic relations which they express among the constituents of the clauses in which they are used.[6] But clearly there is something wrong with a grammatical model, or with an interpretation of a grammatical model,[7] which requires there to be two *hit*'s and three *break*'s. We shall reject this formulation, then, and propose a different description of the facts we have encountered. We shall assume that in some way certain noun phrases can be designated in the deep structure as having an agentive relation to the verbs they are in construction with, others designated as having an instrumental role. Let us label these noun phrases *agent* and *instrument*, and let us assume that a part of our specific knowledge about each verb in our language is a knowledge of the kinds of noun phrases (in the sense we are suggesting) that can occur in construction with it. We are forced to abandon the notions of deep structure subject and deep structure object, if we take this approach, and we must therefore accept a model of grammar in which the subjects and objects that we see in surface structures are introduced by rules.

Temporarily giving the noun phrase *the stick* in examples (5) through (8) the label "X," we may describe *break* as a verb which *requires* an X and which *permits* either an *agent* or an *instrument* or both. Syntactic rules will specify

6. The nature of these differences between our two verbs and the necessary complexity of their description in grammars which require underlying subjects for all sentences are discussed in detail in Hall (1965).

7. In this essay I have considered only interpretations of (5) through (10) that regard them as simplex sentences. A competing view very convincingly presented in Lakoff (1967) would have them *all* be complex (involving successive embeddings) and would assign to them deep structures of an extremely abstract character. I know at the moment no empirical data that would be relevant to a choice between these two views, and it seems to me that neither the facts about *hit* and *break* nor the concluding message about the role of a semantic theory would be affected by a choice between them.

that if there is only an X, the X noun phrase must be the subject. If there is an *agent*, then the X appears as the direct object and the *agent* as subject.[8] If an instrument is selected with *break*, it becomes the subject just in case there is no *agent*, but it shows up at the end of the clause, with the preposition *with*, just in case the sentence does contain an *agent*. The selectional possibilities for *break* can be summarized by saying that it can occur in construction with any of the combinations of noun phrases representable by formula (13):

(13) (*agent*) (*instrument*) X

The parenthesis notation means that the formula holds whether a parenthesized clement is present or not. The left-to-right order of the elements is irrelevant.

By assigning the change of state asserted by *break* to the entity identified by the X noun phrase, by allowing the agentive or instrumental roles of noun phrases to be specified directly by the categories *agent* and *instrument*, by providing for the selection of subjects and objects in the ways mentioned above, and by adopting some formalism which guarantees that noun phrases occurring as *agents* are animate while noun phrases occurring as *instruments* are inanimate, we can account for all of the syntactic and semantic observations that were presented in connection with examples (5) through (8). It is important to realize that this interpretation does not require the separation of *break* into three distinct verbs.

By using the same concepts and rules as those we have just proposed, we can similarly simplify our description of *hit*. We need to say of *hit*, however, that it cannot occur alone with its X element, because sentence (12) is to be excluded. The phenomena we have observed about *hit* can be represented as formula (14)

(14) (*agent*)(*instrument*) X

in which the linked parenthesis notation means that at least one of the two elements so linked *must* be present.

Of course, this "simplification" would be no simplification at all if the categories and rules, and the grammatical distinctions we must recognize to make the rules work, were applicable only to the verbs *hit* and *break*. I have

8. We ignore here the rules that result in "passive" sentences. A more detailed description of the operation of syntactic rules in a grammar of the type suggested here may be seen in Fillmore (1968).

developed elsewhere (Fillmore 1968) the outlines of a general grammatical theory which incorporates modifications of the type we have been discussing. These matters need not concern us here, but it should at least be pointed out that the observations we have made about *break* and *hit* are true of many other English verbs as well. Verbs which are semantically similar to *break* and whose occurrence in clause types is accounted for by formula (13) are exemplified in (15); some English verbs sharing with *hit* properties identified by formula (14) are given in (16).

(15) bend, fold, shatter, crack
(16) slap, strike, bump, stroke

Since (13) and (14) identify "classes" of verbs, it may be the case that certain properties of *hit* and *break* are associated in general with the verb classes to which they belong, other properties being more uniquely associated with the two words as individual lexical items.

In determining what these shared properties might be, we may first note that all of the verbs we chose to associate with *break* assert that the object identified by the X element is understood as undergoing some kind of change of state. That is, the X element is understood as essentially different after the event symbolized by the verb has "happened" to it. But this does not seem to hold for the verbs classified by formula (14). For the purposes of this essay, we shall refer to verbs like *break* and *bend* as "change-of-state" verbs, and verbs like hit and slap as "surface-contact" verbs. The surface-contact verbs assert the occurrence of some physical contact between two objects, but from the use of these verbs one cannot necessarily infer that the objects have undergone any essential change.[9] The most direct way of seeing this fact is by comparing our acceptance of sentence pairs like (17) and (18), in which verbs of these two kinds are contrasted.

(17) I *hit* the window with a hammer; it didn't faze the window, but the hammer shattered.
(18) *I *broke* the window with a hammer; it didn't faze the window, but the hammer shattered.

9. Of course, the surface-contact verbs can also be said to identify a "change of state" of some kind. In a purely abstract sense, a cheek which has once been slapped is different from the same cheek before the slapping event took place. The semantic structures of *some* words recognize properties of objects discoverable not in the objects themselves but in their "histories" (words like *bastard* or *widow*), but such matters nave no relevance to the distinction between the two kinds of verbs we are discussing here.

There is, then, a semantic as well as a syntactic difference between our two classes of verbs. We can capture some of these facts by replacing our temporary symbol X by *place* in formula (14), which we associated with the surface-contact verbs. For the other X we may use (for want of a better word) the term *object*. Now we can reformulate (13) and (14) as (19) and (20), respectively, and associate part of the meaning of expressions containing our verbs with these newly introduced categories of noun phrases.

(19) (*agent*) (*instrument*) *object*
(20) (*agent*)(*instrument*) *place*

The lexical entries for break and hit are assumed to contain references to formulas (19) and (20) respectively.

Change-of-state verbs, as we have said, are verbs which assert of an object a change in time from one "state" to another. An additional syntactic difference between the two verb classes is that "stative adjectives" can be derived from the change-of-state verbs, but not from the others. These adjectives describe the latter of the two states referred to by their underlying verbs. A consequence of this fact is that sentences like those in (21) are ambiguous in ways in which sentences like (22) are not:

(21) $$\text{The window was} \left\{\begin{array}{l}\text{broken}\\ \text{bent}\\ \text{shattered}\end{array}\right\}.$$

(22) $$\text{The window was} \left\{\begin{array}{l}\text{hit}\\ \text{struck}\\ \text{slapped}\end{array}\right\}.$$

The sentences of (21) may be understood either as passives or as descriptions of states, while those of (22) can be understood only as passives.

One more syntactic difference between change-of-state verbs and surface. contact verbs can be seen when the *object* or *place* noun phrase is a possessed body-part noun. The sentences with surface-contact verbs have paraphrases in which the "possessor" appears as the direct object and the body-part noun appears in a "locative prepositional phase."[10]

Notice (23) through (26):

(23) I {hit / slapped / struck} his leg.

(24) I {hit / slapped / struck} him on the leg.

(25) I {broke / bent / shattered} his leg.

(26) * I {broke / bent / shattered} him on the leg.

If the sentences paired by (23) and (24) are correctly interpreted as paraphrases of each other, and if that means that they are identical in their deep structure, it follows that our investigation into the classes of verbs associated with *hit* and *break* requires an understanding of the precise ways in which English grammar deals with body-part nouns.[11]

---

10. This fact, incidentally, tends to lend support to our choice of the category place for the "direct objects" of surface-contact verbs. Some surface-contact verbs, furthermore, permit either the *place* or the *instrument* to become the direct object (where there is an *agent* to serve as subject), allowing such paraphrases as

(i) I hit the roof with the stick
(ii) I hit the stick on the roof.

It should be noticed that when the *instrument* noun phrase is made the direct object, the *place* preposition shows up, and that when the *place* noun phrase is made the direct object, the *instrument* preposition shows up. This is because the rule for forming direct objects, like the rule for forming subjects, has the effect of *deleting the preposition* that would otherwise be associated with the category *agent, place, instrument,* etc.

11. One suggestion on the way in which body-part nouns are to be treated in a grammar is found in the section entitled, "The Grammar of Inalienable Possession," in Fillmore (1968).

I suggested above that the categories *agent* and *instrument* were to be used somehow to guarantee that the noun phrases that filled these positions in sentences would be animate and inanimate respectively. Perhaps a more satisfactory way of dealing with the same facts is to say that the categories *agent* and *instrument* impose a particular interpretation on the nouns that occur "under" them. The sentences we rejected above were sentences in which interpretations imposed on a noun are contradicted by facts that we know about the objects identified by the noun itself. In sentences (8) and (11), for example, we are forced to interpret *A rock* agentively; but since a rock is not animate, and is therefore known to be incapable of initiating any action, we must either interpret the sentences as meaningless, or as fairy-tale sentences in which *A rock* was "personified" (or perhaps the word we need is "animated"). It must remain an open question just what the best treatment of the distinction between normal and semantically anomalous sentences is, but at least we can be clear about the facts. I shall suggest below that my first formulation is more adequate.[12]

Since we have assigned two different categories to the noun phrases we originally labeled X, we must now ask whether the nouns that occur in the X role with these two verbs must be different (or, alternatively, if nouns are to be interpreted in two different ways depending on whether they are identified as *places* or as *objects*).

Consider, in this regard, sentences (27) and (28):

(27) I broke the top of the table.
(28) I hit the top of the table.

In (27), the noun *top* must be referring to the top of a table as a more or less distinct object, while in (28), it can refer either to that or to a portion of the surface area of the table.

This distinction can probably be made clearer with a different kind of example. Suppose you didn't know what *twarge* meant, and you were told two things about some twarge:

(29) John hit the left side of the twarge.
(30) John hit the top of the twarge.

12. For a careful discussion of this difference (or rather, a similar difference), see D. T. Langendoen (1967b).

You might imagine, on hearing (29) and (30), that a twarge was some kind of solid object, and certainly nothing in the two sentences would contradict that assumption. But suppose you were then told two more things about this twarge:

(31) John broke the left side of the twarge.
(32) John broke the top of the twarge.

This time you would be forced to interpret *side* and *top*, not as words designating portions of the surface area of the twarge, but as more or less separable parts of it. The ideas you could have about what a twarge looked like would he much more constrained after you heard (31) and (32) than before. It is clear, in other words, that the X nouns that can occur with *hit* must be partly different (or differently interpretable) from those which occur with *break*.

The fact that in sentences (28) and (30) the noun *top* may be understood as referring *either* to an object or to a location suggests to me that one of the roles of the symbols *agent*, *place*, etc., is in the "selection" of nouns, and that they are not restricted in their effect to the "imposition of an interpretation" on nouns. To appear as an *object*, a noun must identify something that can be a physical object. To appear as a *place*, a noun must identify something that can either be an object or a location.[13]

The ways in which we can understand the nouns that occur with *hit* are almost entirely accounted for by referring to the category *place*, but those connected with the change-of-state verbs appear to involve idiosyncratic properties of individual verbs. In other words, the kinds of objects that *twarge* might refer to are quite different for the various verbs in (34), but not so for these in (33):

(33) I {hit / slapped / struck / smote} the twarge.

13. The distinction could be made more forcefully if we would find nouns which can occur as *places* but never as *objects*. The word *lap* might possibly be such a noun, but I am not sure. At least I am surer about the unacceptability of * *I broke his lap* than I am about that of *I hit his lap*.

(34) I $\left\{\begin{array}{l}\text{bent}\\\text{folded}\\\text{shattered}\\\text{broke}\end{array}\right\}$ the twarge.

Here for the first time our observations become lexically specific. We must now consider what there is to say about objects concerning which it is appropriate to use the word *break*. We have seen already that the object must be in some sense "separable" or discontinuous with other things, but that aspect of the meaning is perhaps best associated with the category *object*.

A sentence like (35)

(35) I broke the dog.

can be interpreted as referring to something which happened to a figurine in the shape of a dog, or perhaps to a frozen dog, but not to an ordinary dog. That is because *break* requires of the entity named by its *object* noun that it be "rigid" in some of its dimensions. One can *break* a dog's bone, but not, ordinarily, a dog.

To show that we are here dealing with a property of the word *break*, rather than with change-of-state verbs in general, we may compare *break* with *bend*. An object that can be *bent* must be "rigid" to the extent that it offers resistance (one doesn't *bend* a handkerchief, normally), but it must also be flexible. Then too, there are more constraints on the permitted "shapes" of objects that can be *bent* than for objects that can be *broken*; but here the best I can do is expect the reader to know what I mean.[14]

There are, then, these more or less specific, and at least intuitively graspable, properties of objects named by nouns capable of appearing nonanomalously with the verb *break*. One could make these observations *seem* more formal, of course, by writing "rigidity" with an initial capital letter and postulating it as a semantic feature of certain nouns, but I believe it would be quite misleading to do so. It seems very unlikely to me that anything is gained by treating these particular "selectional" properties of *break* (and *bend*, *fold*, etc.) in terms of semantic features that are assignable in any natural

14. Notice, too, that *bend* differs from *fold* in respect to both kinds of properties, resistance and shape. The resistance required when one is correctly using the word *bend* is not required for fold, and there are further constraints still or, the shape of objects that can be *folded*. (Once cannot, for instance, *fold* a string.

way to "other words." It is rather that the verb "presupposes" that the "real world objects" named by the nouns that occur with it have certain "physical properties."

But we have still said nothing about what *break* "means." By comparing *break* with *bend* we can see that the former implies, while the latter does not, the appearance of some discontinuity in an external or internal surface of the object; but if there is more to say than that, it is something that can be said as well by dictionary makers as by linguists.

I am inclined to think that the systematic study of the semantic structure of these words ends pretty much with what we have already noted. To seek critical differences between *break, smash, shatter,* to say nothing of the attempt to discover what distinguishes *hit, strike, slam, smite, bump,* etc., from each other, is to involve oneself in judgments that may vary widely from person to person and that may individually have nothing to do with *other* facts about the English language.

Our findings can be summarized by noting that some of the syntactic and semantic properties of our two words can be blamed on the fact that they are words of a particular type; that is, many of the facts we have encountered are instances of general facts about whole classes of words or about uses and interpretations of grammatical categories the existence of which is determinable independently of questions of lexicography.

Both of these verbs can be used transitively, with subjects that are understood agentively as well as with subjects that are understood instrumentally. But *break,* unlike *hit,* can also be used intransitively. *Break* is one of a class of verbs used to express a change of state in some object, and as such it provides, in the form of its past participle, a stative adjective which is capable of describing the object in its latter state. One of the properties of *hit,* namely the existence of certain kinds of paraphrases when it is used with body-part nouns, is apparently accounted for by referring to the ways in which body-part nouns are used as indicators of places rather than as indicators of objects. And constraints on the specific nouns that can occur in constructions with these verbs are partly determined by the categories *agent, instrument, place,* and *object.*

All of these phenomena are either ascribable to larger grammatical facts or to whole classes of verbs. They fall under what really amounts to our "general" knowledge of English, and are therefore to be treated only in the grammar. The only word-specific information that is related to the preceding observations is: (a) that given in formulas (19) and (20), the information that indicates, for each of these verbs, the syntactic environments in which it is appropriate to use it; and (b) the information that *break* semantically expresses a change of state.

Apart from the information about their general semantic character (as change-of-state verbs or not) and the needed indication of the environments into which they can be inserted, the only really specific lexical information that we have encountered are: (a) the special ways in which speakers of English accept the *object* nouns that occur with *break* and interpret them; and (b) the specific meaning of the two words. The word *break* can be appropriately used only with an object that is "rigid" in some of its dimensions, and it expresses the appearance of some discontinuity therein. (But why one can *break* a thread but not a cloth is not easily covered by this statement.) The nouns that can occur with *hit* apparently do not need to satisfy any requirements not associated with their categorization as *places*, and what exactly is meant by *hit*, in the sense of the kind of surface contact asserted by *hit* in particular (as opposed to *strike*, etc.), is extremely difficult to pin down.

Not every change-of-state verb can be used intransitively: consider *smash* and *cut* in this respect. Not every verb having syntactic properties of the kind represented by formula (19) is a change-of-state verb: there is a large class of motion verbs syntactically like *break*, including *move*, *turn*, *rotate*, *spin*, etc. Not every verb that is semantically a change-of-state verb and is syntactically transitive or intransitive in a way analogous to that represented by formula (19) takes an *object*: some take animate nouns, such as *wake* (*up*). Some change-of-state verbs are understood as affecting a place on an object rather than an object as a whole. The verbs *cut* and *bite*, for example, are of this type, and show paraphrase relations of the type seen in (25) and (26) with body-part nouns. And some change-of-state verbs have associated adjectives that are not identical in form with their past participles. The word *awake* is the stative adjective for *wake* (*up*), and that must have something to do with the fact that nobody knows what the past participle of *wake* is!

What these observations show is that many of the apparent regularitics suggested by this study are slightly spurious, and what that means is that the lexical description of some of the other verbs we have mentioned will be somewhat more detailed than that of *break* or *hit*.

One of the conclusions that can be drawn from this study is that the data we have examined fail to support the distinction between syntax (as providing a "level" of representation) and semantics. Modifications that are being devised in the theory of deep structure are tending more and more to provide concepts of the kind that can be used quite directly for expressing semantic assertions about linguistic expressions. The designation of noun phrases as *agents*, *places*, etc., that has been used in my approach, has a role in semantic interpretation, and such properties of verbs as the change-of-state feature we have associated with *break* are semantic in a more obvious way.

The assignment of such semantic features has, however, clear "syntactic consequences." Current developments within the theory of generative transformational grammar suggest that all "purely syntactic" concepts in grammar relate to the application of syntactic rules, not to the underlying representation of sentences.[15]

Certain other matters that have been considered proper to semantics but distinct from syntax include formalizations of the notion of semantic anomaly through semantic projection rules of the type proposed by Katz and Fodor (1963). These authors distinguish between "markers" and "distinguishers" among the features that compose semantic characterizations of lexical items. The distinguishers are those features of the semantic description of a word which are idiosyncratic to that word. The markers are those features which enter into semantic generalizations, features in terms of which various semantic judgments on sentences–including judgments on semantic anomaly–can be formalized and made explicit.

Since a part of the description of certain words is a statement of their selectional restrictions–that is, a statement of the conditions that determine their meaningful use–it should be the case that all semantic anomalies should be describable by a single device. Our consideration of the conditions of appropriateness for the use of words like *break*, *bend*, and *fold*, however, suggests that selectional information can be as idiosyncratic as the kinds of properties that have been referred to as semantic distinguishers. The treatment of the selectional properties of verbs cannot be carried out in a non-*ad hoc* way, it seems to me, by seeking features on nouns which do or do not violate restrictions associated with particular change-of-state verbs. In fact, it looks very much as if for a considerable portion of the vocabulary of a language, the conditions determining the appropriate use of a word involve statements about properties of real world objects rather than statements about the semantic features of words.

Some facts about language that have been hitherto treated in terms of a

15. I have in mind unpublished manuscripts by J. D. McCawley, J. R. Ross, G. Lakoff, and D. T. Langendoen.

semantic interpretive component viewed as distinct from the syntactic component have been absorbed into the latter–that is, they have been shown lo he explainable within a combined syntactic–semantic component. Other facts that have been treated by some as belonging to semantic theory proper are believed to be more correctly assigned to the study of the speakers' "practical" knowledge of their language. It seems to me that the explanatory scope of semantics as such, to the extent that semantic knowledge can be separated from knowledge of syntax (or syntax–semantics) and knowledge of the world, should be limited to a clarification of the conceptual interrelatedness of lexical items and the semantic judgments on sentences that can be directly accounted for in terms of this interrelatedness.

4

Types of Lexical Information

This paper attempts a survey of the kinds of information needed in a semantically and syntactically complete lexicon. I begin with introducing an "extra-propositional" layer of relevant role types, namely those involving the parameters of the communication event within which sentences (and hence their words) occur: the place and time of the event, and the communicator and the addressee. Of special interest are words or other linguistic formations that relate to articulations between aspects of what can be referred to as the reporting event *and the* reported eventuality. *The terminology introduced here, such as PLS and PLT–'place of the locutionary source' (where I am) and 'place of the locutionary target' (where you are)–was never used again, at least not by me. The attempt here was to incorporate some of Haj Ross's observations on interactions between word meanings and appropriateness judgments about reporting participants. (In an utterance like* John misunderstands the situation *the speaker claims that John has some understanding of the situation and that understanding is wrong; an utterance of the simple present-tense sentence* I misunderstand the situation *is bizarre precisely because it claims something contradictory of the speaker: I have an understanding of the situation–from the fact that the first-person pronoun is the subject of the verb–and that understanding is wrong–from the fact that the speaker is the locutionary source. Tense, aspect and modality interact with such things in various ways that are explained by the double-barrel interpretation of a verb like* misunderstand*: it is all right to say* I have misunderstood the situation, I misunderstood the situation, perhaps I am misunderstanding the situation, *and so on, but not the unmodified first-person present-tense version.)*

A proposal that was new within case theory was the suggestion that more than one case can be assigned to a single constituent. Where verbs like lift *and* raise *call for separate realization of the lifting agent and the lifted entity, requiring reflexive syntax if the two individuals are identical,* arise *identifies the agent with the ascending entity, whereas* rise *and* ascend *do not. (Concepts like* uner-

Originally published in 1969

gative *and* unaccusative *had not yet been introduced to the world.) The verbs* sell *and* buy *can similarly be distinguished by associating the Agent with the Source of the motion (of the merchandise) in the case of* sell *and the Goal of the motion in the case of* buy.

Questions of the number of "arguments" of a predicate require some kind of story about optionality versus obligatoriness, and that requires an account of the difference between being conceptually necessary *and being syntactically necessary. The transitive and intransitive uses of* break *should not be spoken of in terms of simple optionality of expression:* break *with an agent present has a different meaning than otherwise. We need to consider cases in which elements that are inescapably a part of a situation are nevertheless omissible in sentences that evoke that situation; but for these omissible elements we need now to distinguish two kinds, those where the omitted information is simply not shared between speaker and hearer, and those where the omitted element is "understood" in the context shared by speaker and hearer. Thirdly, we have elements that are not separately mentioned because they are "incorporated" into the meaning of a verb: the verb* spank *usually occurs without explicit mention of either the weapon or the target.*

The article introduced a kind of template for lexical entries that was intended to provide a place for all of the kinds of information outlined in the article, and a few examples are offered that include complementation requirements, preposition selection, single versus multiple role assignments to syntactic arguments, the conditions governing argument omission, distinctions between what is asserted and what is presupposed in sentences headed by given predicates, and a few other things. Much of my own work since then has been aimed at completing a notion of the structure of lexical entries in a scientific lexicon.

0. The Lexicon

A lexicon viewed as part of the apparatus of a generative grammar must make accessible to its users, for each lexical item,

(i) the nature of the deep-structure syntactic environments into which the item may be inserted;

(ii) the properties of the item to which the rules of grammar are sensitive;

(iii) for an item that can be used as a 'predicate', the number of 'arguments' that it conceptually requires;

(iv) the role(s) which each argument plays in the situation which the item, as predicate, can be used to indicate;

(v) the presuppositions or 'happiness conditions' for the use of the item, the conditions which must be satisfied in order for the item to be used 'aptly';

(vi) the nature of the conceptual or morphological relatedness of the item to other items in the lexicon;

(vii) its meaning; and

(viii) the phonological or orthographic shapes which the item assumes under given grammatical conditions.

In this paper I shall survey, in a very informal manner, the various types of information that needs to be included, in one way or another, in the lexical component of an adequate grammar. I shall, however, have nothing to say about (viii) above, and nothing very reliable to say about (vii).

1. The Speech Act

I begin by assuming that the semantic description of lexical items capable of functioning as predicates can be expressed as complex statements about properties of, changes in, or relations between entities of the following two sorts: (a) the entities that can serve as arguments in the predicate-argument constructions in which the given lexical item can figure, and (b)various aspects of the speech act itself. In this section I shall deal with the concepts that appear to be necessary for identifying the speech act in semantic theory.

The act of producing a linguistic utterance in a particular situation involves a speaker, an addressee, and a message. It is an act, furthermore, which occurs within a specific time-span, and it is one in which the participants are situated in particular places. Now the time during which a speech act is produced is a span, the participants in the speech act may be moving about during this span, and even the identity of the participants may change during the speech act; but for most purposes the participant-identity and the time-space coordinates of the speech act can be thought of as fixed points. In accepting this fiction, I commit myself to regarding sentences like 1 to 3 as somewhat pathological:

(1) I'm not talking to you, I'm talking to you.
(2) I want you to turn the corner... right... here!
(3) This won't take long, did it?

The producer of a speech act will be called the *locutionary source* (LS), the addressee will be referred to as the *locutionary target* (LT)[1]. The temporal and

1. I have borrowed these terms from philosopher Richard Garner, to whom I am also indebted for a number of suggestions on the content and phrasing of several sections of this paper.

spatial coordinates of the speech act are the *time of the locutionary act* (TLA), the *place of the locutionary source* (PLS) and the *place of the locutionary target* (PLT).

There are certain verbs in English which refer to instances of speech acts other than the one which is being performed (e.g. *say*), and I shall refer to these as *locutionary verbs*. It is necessary to mention locutionary verbs now because we shall find that linguistic theory requires a distinction between the 'ultimate' speech act and speech acts described or referred to in a sentence. Thus in the linguistic description of some verbs reference is made to either the LS or the agent of a locutionary verb[2], and these situations must be distinguished from those in which the reference is to the LS alone.

Words in English whose semantic descriptions require reference to some aspect of the locutionary act include *here, this, now, today, come,* and *know.* The word *come*[3], for example, can refer to movement toward either the PLA or the PLT at either TLA or the time-of-focus identified in the sentence. Thus in 4

(4) He said that she would come to the office Thursday morning

it is understood that the office is the location of the LS or the LT either at TLA or on the said Thursday morning. Uses of the verb *know* allow the LT to infer the factuality of the proposition represented by a following *that*-clause. Thus, in sentence 5

(5) She knows that her brother has resigned

it is understood that the LS at TLA presupposes the factuality of her brother's resignation.

In the semantic description of some verbs reference is made either to the subject of the 'next locutionary verb up', or to the LS just in case the sentence contains no explicit locutionary verb. If we are to believe Ross (see note 2), the verb *lurk* requires of its subject that it be distinct from the subject of the first commanding locutionary verb. It may look, on just seeing sentences 6–7,

(6) He was lurking outside her window

2. The disjunction in this statement may be unnecessary if we accept John R. Ross's arguments that declarative sentences have phonetically unrealized embedding sentences representable as something like *I declare to you that....* On Ross's view every sentence contains at least one locutionary verb, so that the difference we are after is a difference between references to the 'next higher' locutionary verb and reference to the 'highest' locutionary verbs (Ross 1969).

3. See Fillmore (1966b).

(7) *I was lurking outside her window

that what is required is simply non-identity with LS; but this is shown not to be so because of the acceptability of sentence 8

(8) She said I had been lurking outside her window.

From these observations it follows that sentence 9 is ambiguous on whether or not the two pronouns *he* are coreferential, but sentence 10 requires the two *he*s to be different.

(9) He said he had been loitering outside her window
(10) He said he had been lurking outside her window.

There are apparently many speakers of English whose use of *lurk* fails to match the observations I have just reviewed; but for the remainder, this verb provides an example of the distinction we are after.

There is, then, a distinction in semantic descriptions of lexical items between references to properties of the higher clauses that contain them, on the one hand, and to features of or participants in the speech act itself on the other hand. The former situation falls within the area of 'deep structure constraints' (see Section 9 below), but the latter requires the availability of concepts related to the speech act.[4]

4. There is an additional use of deictic words, and that is this: in a third person narrative, one can express one's 'identification' with one of the characters in the narrative by letting that character be the focus of words that are primarily appropriate to *hic-nunc-ego*. Thus we may find in an exclusively third-person narrative a passage like i

(i) here was where Francis had always hoped to be, and today was to mark for him the beginning of a new life

in which the words *here* and *today* refer to the place and time focused on in the narrative, not to the place and time associated with the author's act of communication. In what might be referred to as the 'displaced ego' use of deictic words, the author has shown us that he has for the moment assumed Francis's point of view.

I propose that a rather subtle test of psycho-sexual identity can be devised which makes use of a story in which two characters, one male and one female, do a lot of cross-visiting, but in their other activities do nothing that makes one of them clearly more lovable than the other. The subject's task is to listen to the story and then retell it in his own words. The writer of the original story must not use the words *come* and *go*; but if the subject, in retelling the story, states, say, that *Bill came to Mary's house* (using *came* rather than *went*), this fact will reveal that he is experiencing the story from Mary's point of view.

2. Elementary Semantic Properties Of Verbs

I assume that what we might call the basic sense of a word is typically expressible as a set of components, and that while some of these components may be idiosyncratic to particular words, others are common to possibly quite large classes of words. The components themselves may be complex, since they may be required to characterize events or situations that are themselves complex, but the ultimate terms of a semantic description I take to be such presumably biologically given notions as identity, time, space, body, movement, territory, life, fear, etc., as well as undefined terms that directly identify aspects of or objects in the cultural and physical universe in which human beings live. In this section, we shall sample certain elementary semantic properties of verbs, in particular those relating to time, space, movement, and 'will'.

Some verbs refer to activities viewed as necessarily changing in time, others do not; this contrast is frequently referred to with the terms 'momentary' and 'continuative' respectively. *Sleep* is a continuative verb, *wake up* is a momentary verb. A continuing activity, or state, necessarily occupies a span of time, and thus it makes sense to qualify a continuative verb with a complement which identified one or both of the end-points of such a span, or a distance-measure of the span. Thus while the sentences in 11 make sense, those in 12 do not.

(11) She slept $\left\{\begin{array}{l}\text{for three days}\\\text{until friday}\end{array}\right\}$

(12) *She woke up $\left\{\begin{array}{l}\text{for three days}\\\text{until friday}\end{array}\right\}$

On the other hand, the negation of a momentary verb can identify a continuing state; hence, the sentences in 13 make sense.

(13) She didn't wake up $\left\{\begin{array}{l}\text{for three days}\\\text{until friday}\end{array}\right\}$

Momentary verbs that represent acts that are repeatable may be understood 'iteratively'. *Wake up* is not iterative, as is shown by example 12. *Kick*, however, can be understood iteratively, as we see in 14

(14) He kicked the dog until 5 o'clock.

Momentary verbs that are also 'change-of-state' verbs cannot be used iteratively when a specific object is involved, as we see in example 15,

(15) *He broke the vase until 5 o'clock

but if the same activity can be directed to an unspecified number of objects, then change-of-state verbs can be understood iteratively, too, as we see in 16

(16) He broke vases until 5 o'clock.

It appears, in short, that a lexico-semantic theory will have to deal with those aspects of the meanings of verbs which relate to the occurrence in time of the situations which they identify.

Turning to other types of semantic properties, we note that verbs like *hit* and *touch*, though they differ in that the former is momentary while the latter may be either momentary or continuative, have in common the notion of surface-contact, a property they share with *knock, strike, contact, impinge, smite*, and many others. They differ in that the impact of the described acts is apparently gentler for *touch* than for *hit*.

The verbs *leap* and *jump* agree in implying a momentary change in vertical position (one has to leave the ground in order to perform either of these actions), but they differ in that *leap* seems to imply a change in horizontal position, too. *Slide*, like *leap*, refers to position-changes along a surface, but differs from *leap* in not implying movement away from the surface. *Scuttle*, like *slide*, suggests movement across a surface, but with the assumption that contact with the surface is interrupted and with the further sense that the motion is rapid. *Dart* is like *scuttle* in referring to rapid sudden motion, but fails to share with it any reference to a surface. Verbs of motion, in short, may be described by associating them with properties relating to direction, speed, gravity, surface, etc.[5]

Sometimes a verb has a built-in reference to the outcome of an activity. Conceptually it appears that the actor engages in some activity and though the activity may be directed toward some specific outcome it is the activity itself which (by chance) leads to that outcome. These have been called 'achievement verbs'. One of the tests of an achievement verb is that the modal

5. Possibly the richest source of insights into verbs of motion is the recent output of Jeffrey S. Gruber (1965, 1967a, 1967).

may is usable in construction with such a verb only in its epistemic or predictive sense, not in its pragmatic or permission-granting sense. This is apparently because of the 'by-chance' relationship between the activity and the outcome: one doesn't grant someone permission to have good luck. Hence we find 17 and 19 understandable only in the epistemic sense of *may*, while 18 and 20 can be understood in either the epistemic or the pragmatic sense.

(17) He may achieve his goal
(18) He may try to achieve his goal
(19) He may find the eggs
(20) He may look for the eggs.

A final general property of verbs that we may point out in this section has to do with the intentional or non-intentional involvement of one of the participants in the events described by use of the verb. If we compare 21 and 22

(21) John means *x* by *y*
(22) John understands *x* by *y*

we note that in 21, but not in 22, the association between *x* and *y* is intentional on John's part. The word *mean* can be used in the sense which *understand* has in 22, but in that case the sentence is differently constructed. Sentence 23 is a paraphrase of 22.

(23) *Y* means *x* to John.

3. Predicate Structure

I assume that most of the 'content words' in a language can be characterized in the lexicon in terms of their use as predicates. I take this to be true of nouns, verbs, adjectives[6], most adverbs, and also a great many conjunctions. Thus a sentence like 24

6. In other words, I accept the part-of-speech identities argued by George Lakoff in Appendix A of *On the Nature of Syntactic Irregularity*, Report No. NSF-16, Computation Laboratory of Harvard University, 1965; as well as the extension of such identities to 'nouns' proposed by Emmon Bach in 'Nouns and Noun Phrases', in *Universals in Linguistic Theory* (ed. by E. Bach and R. Harms), Holt, Rinehart and Winston, 1968.

(24) Harry lives at home because he loves his mother

is evaluated as true or false depending not only on the joint truth-values of the two clauses which flank *because*, but on the truth or falsity of the 'causal' connection between the two situations named by these clauses. The sentence can be interpreted as having *because* as its main predicate, a predicate which takes two clauses as its arguments and which is used to assert a 'causal' or 'logical' connection between them.

As predicates, words can be described first of all according to the number of 'arguments' that they take. Thus the verbs *ascend* and *lift* are both motion verbs, they are both used to describe motion upward, but they differ in that while *ascend* is used only of the object that moves upward, *lift* requires conceptually two objects, one the object that is moving upward, the other the object or being that is causing it to move upward. Another way of stating this is: *ascend* is a one-argument predicate, *lift* is a two-argument predicate.[7]

Many verbs are flexible in the number of arguments they take. This is true, for example, of some motion verbs, like *move* and *rotate*, and many change-of-state verbs, like *open* and *break*. *Move*, as can be seen in sentences 25–27, can occur with one, two, or three arguments.

(25) The rock moved
(26) The wind moved the rock
(27) I moved the rock (with a stick).

Mention of the object which moves is required of all three uses; the two-argument uses additionally identify either the physical force or object which is directly responsible, or the animate being which is indirectly responsible, for the activity of moving; and the three-argument use identified all three of these (as in 27 with the parenthesized phrase included). The surface-contact verbs *hit, touch, strike*, etc., require conceptually at least two arguments in all of their uses, namely the objects which come into contact, but they accept as a third argument the animate being that is responsible for the coming-into-contact.

The verbs *rob* and *steal* conceptually require three arguments, namely those identifiable as the culprit, the loser, and the loot. The words *buy* and *sell*

7. Of course, as motion verbs each of them may take time and space complements as well, as is seen in *The balloons ascended to the rafters just after the speech ended.* Since in general the nature of the time and space complements is predictable from properties of the type discussed in Section 2, we may permit ourselves to ignore such matters while discussing the typing of predicates on the basis of the number of arguments they accept.

are each 4-argument predicates, the arguments representing the one who receives the goods or services, the one who provides the goods and services, the goods and services themselves, and the sum of money that changes hands.

I have referred in this section to the conceptually required number of arguments. I am distinguishing this from the number of arguments that must be explicitly identified in English sentences. The various ways in which English grammar provides for the omission or suppression of conceptually required arguments is the subject of Section 5. To say that conceptually *rob* or *buy* are 3- or 4-argument predicates respectively is to acknowledge that even when we say merely 28

(28) She robbed the bank

we understand that she took something out of the bank, and when we say 29

(29) She bought it

truthfully, it is necessarily the case that there was somebody who sold it to her and that a sum of money was exchanged.

4. Case Structure

In the preceding section I identified the separate arguments associated with the verb *rob* by referring to their roles as 'culprit', 'loser', and 'loot'; in a similar way, I might have identified the three arguments associated with the verb *criticize* as 'critic', 'offender' and 'offense'. It seems to me, however, that this sort of detail is unnecessary, and that what we need are abstractions from these specific role descriptions, abstractions which will allow us to recognize that certain elementary role notions recur in many situations, and which will allow us to acknowledge that differences in detail between partly similar roles are due to differences in the meanings of the associated verbs. Thus we can identify the culprit of *rob* and the critic of *criticize* with the more abstract role of Agent, and interpret the term Agent as referring, wherever it occurs, as the animate instigator of events referred to by the associated verb. Although there are many substantive difficulties in determining the role structure of given expressions, in general it seems to me that for the predicates provided in natural languages, the roles that their arguments play are taken from an inventory of role types fixed by grammatical theory. Since the most readily available terms for these roles are found in the literature of case theory, I have taken to referring to the roles as *case relationships*, or simple *cases*. The combination of cases that might be associated with a given predicate is the *case structure* of that predicate.

In addition to the apparently quite complex collection of complements that identify the limits and extents in space and time that are required by verbs of motion, location, duration, etc., the case notions that are most relevant to the subclassification of verb types include the following:

Agent (A), the instigator of the event
Counter-Agent (C), the force or resistance against which the action is carried out
Object (O), the entity that moves or changes or whose position or existence is in consideration
Result (R), the entity that comes into existence as a result of the action
Instrument (I), the stimulus or immediate physical cause of an event
Source (S), the place to which something is directed
Experiencer (E), the entity which receives or accepts or experiences or undergoes the effect of an action (earlier called by me 'Dative').

It appears that there is sometimes a one-many relationship between an argument to a predicate and the roles that are associated with it. This can be phrased by saying either that some arguments simultaneously serve in more than one role, or that in some situations the arguments in different roles must (or may) be identical.

Thus verbs like *rise* and *move* can be used intransitively, that is with one noun phrase complement; the complement may refer just to the thing which is moving upward, or it may simultaneously refer to the being responsible for such motion. Thus in speaking simply of the upward motion of smoke we can say 30,

(30) The smoke rose

and in speaking of John's getting up on his own power, we can say

(31) John rose.

The case structure of *rise* then, might be represented diagrammatically as 32. The fact that there are two case lines connecting *rise* with its one argument, and that the line labeled A has its case label in parentheses, reflect the fact

(32)

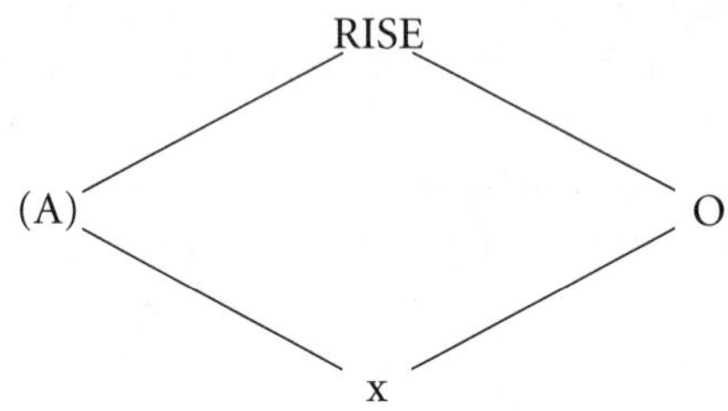

that the argument can serve in just one of these roles (O) or simultaneously in both (A and O). *Rise* differs from *arise* in the optionality of A; it differs from *ascend* in having an 'A' line at all, and they all differ from *lift* in that the latter requires two arguments.[8] Thus 33–35

---

8. In truth, however, *lift* also requires conceptually the notion of Instrument. That is, lifting requires the use of something (perhaps the Agent's hand) to make something go up. It is conceivable that the basic case structures of *arise* and *lift* are identical, with grammatical requirements on identity and deletion accounting for their differences. Thus the two are shown in i and ii.

(i), (ii)

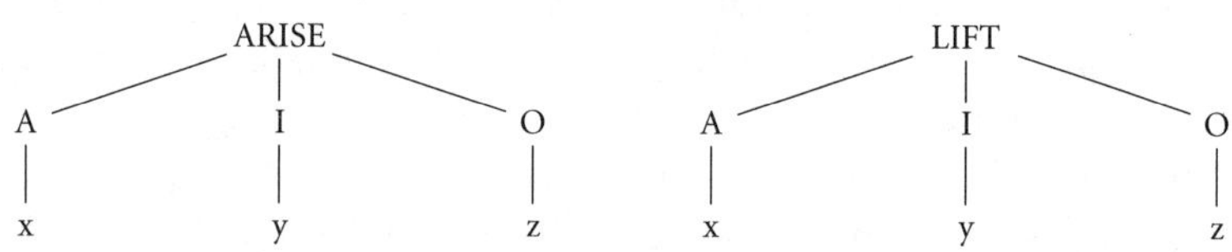

For *arise*, however, it is required that $x = y = z$, and hence only one noun will show up. The meaning expressed by *John arose* is that John willed his getting up, that he used his own body (i.e., its muscles) in getting up, and that it was his body that rose. For *lift*, however, there may be identities between *x* and *z*, resulting in a sentence like *John lifted himself*, or there may be identities between *y* and *z*, resulting in a sentence like *John lifted his finger*. *Rise*, then, must be described as (iii)

(iii)

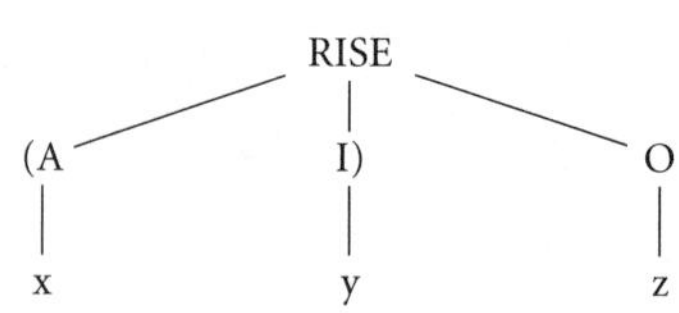

where the identity requirements of *arise* obtain just in case *x* and *y* are present.

(33)

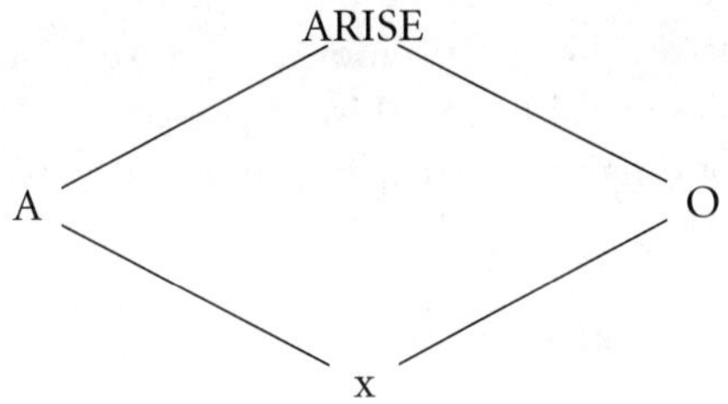

(34)

(35)

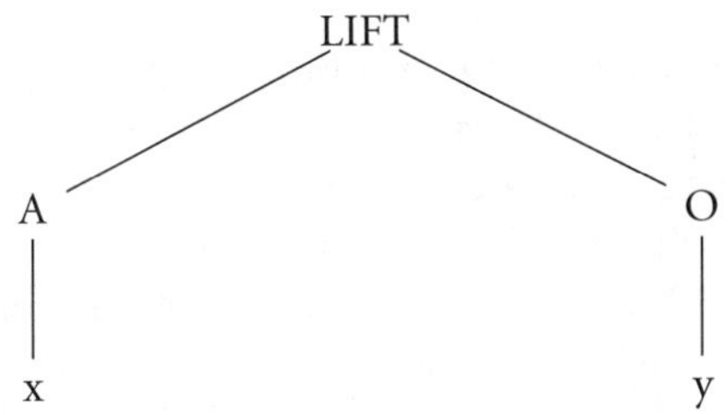

Frequently a linguistically codable event is one which in fact allows more than one individual to be actively or agentively involved. In any given linguistic expression of such an event, however, the Agent role can only be associated with one of these. In such pairs as *buy* and *sell* or *teach* and *learn* we have a Source (of goods or knowledge) and a Goal. When the Source is simultaneously the Agent, one uses *sell* and *teach*; when the Goal is simultaneously the Agent, we use *buy* and *learn*.

It is not true, in other words, that *buy* and *sell*, *teach* and *learn* are simply synonymous verbs that differ from each other in the order in which the arguments are mentioned.[9] There is synonymy in the basic meanings of the verbs (as descriptions of events), but a fact that might be overlooked is that each of these verbs emphasizes the contribution to the event of one of the participants. Since the roles are different, this difference is reflected in the ways in which the actions of different participants in the same event can be qualified. This we can say 36

(36) He bought it with silver money

but not 37

(37) He sold it with silver money.

Similarly, the adverbs in 38 and 39 do not further describe the activity as a whole, but only one person's end of it.

(38) He sells eggs very skilfully
(39) He buys eggs very skilfully.

5. Surface Realization Of Arguments

I suggested in the previous section that the conceptually necessary arguments to a predicate cannot always be matched on a one-to-one basis with the 'cases' that are also associated with the same predicate. It may now be pointed out that there is also no exact correspondence between either of these and the number of obligatorily present syntactic constituents in expressions containing the predicates in question. *Buy*, as we have seen, is a 4-argument (but 5-case) predicate which can occur in syntactically complete sentences containing 2, 3, or 4 noun phrases. Thus, 40,

(40) He bought it (from me) (for four dollars)

in which optionally present segments are marked off by parentheses.

The verb *blame* has associated with it 3 roles, the accuser (Source), the defendant (Goal), and the offense (Object). Expressions with this verb can contain reference to all three, as in 41, just two, as in 42 and 43, or only one, as in 44.

9. This has, of course, been suggested in a great many writings on semantic theory; most recently, perhaps, in the exchanges between J. F. Staal and Yehoshua Bar-Hillel in *Foundations of Language*, 1967 and 1968.

(41) The boys blamed the girls for the mess
(42) The boys blamed the girls
(43) The girls were blamed for the mess
(44) The girls were blamed.

No sentence with *blame*, however, can mention only the accuser, only the offense, or just the accuser and the offense. See 45–47.

(45) *The boys blamed
(46) *The mess was blamed
(47) *The boys blamed (for) the mess.

An examination of 41–47 reveals that the ease realized here as *the girls* is obligatory in all expressions containing this verb, and, importantly, that there are two distinct situations in which the speaker may be silent about one of the other arguments. I take sentence 43 as a syntactically complete sentence, in the sense that it can appropriately initiate a discourse (as long as the addressee knows who the girls are and what the mess is). In this case the speaker is merely being indefinite or non-committal about the identity of the accuser. I take sentence 42, however, as one which cannot initiate a conversation and one which is usable only in a context in which the addressee is in a position to know what it is that the girls are being blamed for. Another way of saying this is that 43 is a paraphrase of 43' and 42 is a paraphrase of 42'.

(42') The boys blamed the girls for it.
(43') The girls were blamed for the mess by someone.

This distinction can be further illustrated with uses of *hit*. In 48, a paraphrase of 48', the speaker is merely being indefinite about the implement he used:

(48) I hit the dog
(48') I hit the dog with something.

In 49, the paraphrase of 49', the speaker expects the identity of the 'target' (Goal) to be already known by the addressee.

(49) The arrow hit
(49') The arrow hit it.

The two situations correspond, in other words, to definite and indefinite pronominalization.

Sometimes an argument is obligatorily left out of the surface structure because it is subsumed as a part of the meaning of the predicate. This situation has been discussed in great detail by Jeffrey Gruber (see note 5 above) under the label 'incorporation'. An example of a verb with an 'incorporated' Object is *dine*, which is conceptually the same as *eat dinner* but which does not tolerate a direct object.[10]

The verb *doff* has an incorporated Source. If I *doff* something, I remove it from my head, but there is no way of expressing the Source when this verb is used. There is no such sentence as 50

(50) *He doffed his hat from his balding head.

There are other verbs that identify events which typically involve an entity of a fairly specific sort, so that to fail to mention the entity is to be understood as intending the usual situation. It is usually clear that an act of slapping is done with open hands, an act of kicking with legs and/or feet, an act of kissing with both lips; and the target of an act of spanking seldom needs to be made explicit. For these verbs, however, if the usually omitted item needs to be delimited or qualified in some way, the entity can be mentioned. Hence we find the modified noun phrase acceptable in such sentences as 51–53

(51) She slapped me with her left hand[11]
(52) She kicked me with her bare foot
(53) She kissed me with chocolate-smeared lips.

Lexical entries for predicate words, as we have seen in this section, should represent information of the following kinds: (1) for certain predicates the nature of one or more of the arguments is taken as part of our understanding of the predicate word: for some of these the argument cannot be given any linguistic expression whatever; for others the argument is linguistically identified only if qualified or quantified in some not fully expected way; (2) for certain predicates, silence ('zero') can replace one of the argument-expressions just in case the speaker wishes to be indefinite or non-commital about the identity of the argument; and (3) for certain predicates, silence can replace

10. One can, however, indicate the content of the meal in question, as in an expression like *He dined on raisins.*

11. At least in the case of *slap*, the action can be carried out with objects other than the usual one. Thus it is perfectly acceptable to say *She slapped me with a fish.*

one of the argument-expressions just in case the LS believes that the identity of the argument is already known by the LT.

6. Meaning vs. Presupposition

Sentences in natural language are used for asking questions, giving commands, making assertions, expressing feelings, etc. In this section I shall deal with a distinction between the presuppositional aspect of the semantic structure of a predicate on the one hand and the 'meaning' proper of the predicate on the other hand. We may identify the presuppositions of a sentence as those conditions which must be satisfied before the sentence can be used in any of the functions just mentioned. Thus the sentence identified as 54

(54) Please open the door

can be used as a command only if the LT is in a position to know what door has been mentioned and only if that door is not, at TLA, open.[12] The test that the existence and specificity of a door and its being in a closed state make up the *presuppositions* of 54 rather than part of its *meaning* is that under negation the sentence is used to give quite different instructions, yet the presuppositional conditions are unaffected.

(54') Please don't open the door.

The presuppositions about the existence and specificity of the door relate to the use of the definite article and have been much discussed in the philosophical literature on referring.[13] The presupposition about the closed state of the door is a property of the verb *open.*

Presuppositions of sentences may be associated with grammatical constructions independent of specific predicate words (such as those associated with the word *even* or with the counterfactual-conditional construction[14]), but I shall mention here only those that must be identified with the semantic

12. I am only dealing here with those presuppositions which are relatable to the *content* of the utterance. It is also true, of course, that 54 can be used appropriately as a command only if the LT understands English, is believed by the LS to be awake, is not totally paralyzed, etc. These matters have more to do with questions of 'good faith' in speech communication than with information that is to be understood as knowledge about individual lexical items.

13. On referring see "On referring", Strawson (1950).

14. For some examples see my mistitled paper, 'Entailment Rules in a Semantic Theory', *Project on Linguistic Analysis* Report No. 10, 1965.

structure of predicate words. If we limit our considerations to sentences which can be used for making assertions, we can separate the basic meaning of a predicate from its presuppositions, by describing the former as being relevant to determining whether as an assertion it is true or false, the latter as being relevant to determining whether the sentence is capable of being an assertion in the first place. If the presuppositional conditions are not satisfied, the sentence is simply not apt; only if these conditions are satisfied can a sentence be appropriately used for asking a question, issuing a command, making an apology, pronouncing a moral or aesthetic judgment, or, in the cases we shall consider, making an assertion.

Let us illustrate the distinction we are after with the verb *prove* in construction with two *that*-clauses. Consider sentence 55

(55) That Harry is still living with his mother proves that he is a bad marriage risk.

It is apparent that if I were to say 55 about somebody who is an orphan, nobody would say that I was speaking falsely, only that I was speaking inappropriately. If *prove* has a *that*-clause subject and a *that*-clause object, we say that the truth of the first *that*-clause is *presupposed*, and that the verb is used to assert a causal or logical connection between the two clauses and thus (when used affirmatively) to imply the truth of the second clause. That this separation is correct may be seen by replacing *prove* in 55 by *doesn't prove* and noting that the presuppositional aspects of 55, concerning the truth of the first *that*-clause, are unaffected by the change.

It is difficult to find pairs of words in which the presuppositional content of one is the meaning content of the other, but a fairly close approximation to this situation is found in the pair of verbs *accuse* and *criticize*. The words differ from each other on other grounds, in that *accuse* is capable of being a 'performative', while *criticize* is not; and *criticize*, unlike *accuse*, is capable of being used in senses where no negative evaluation is intended. In sentences 56 and 57 we are using *accuse* in a non-performative sense and we are using *criticize* as a 3-argument predicate in a 'negative-evaluative' sense.

(56) Harry accused Mary of writing the editorial

(57) Harry criticized Mary for writing the editorial.

I would say that a speaker who utters 56 presupposes that Harry regarded the editorial-writing activity as 'bad' and asserts that Harry claimed that Mary was the one who did it; while a speaker who utters 57 presupposes that Harry regarded Mary as the writer of the editorial and asserts that Harry claimed the editorial-writing behavior or its result as being 'bad'. The content of the presupposition in each one of these verbs shows up in the assertive meaning of the other.

Certain apparent counter-examples to the claims I have been making about presuppositions can be interpreted as 'semi-quotations', I believe. Some utterances are to be thought of as comments on the appropriate use of words. Uses of the verb *chase* presuppose that the entity identified as the direct object is moving fast. Uses of the verb *escape* presuppose that the entity identified by the subject noun phrase was contained somewhere 'by force' previous to the time of focus. These presuppositions, as expected, are unaffected by sentence negation.

(58) The dog $\left\{\begin{array}{l}\text{chased}\\ \text{didn't chase}\end{array}\right\}$ the cat

(59) He $\left\{\begin{array}{l}\text{escaped}\\ \text{didn't escape}\end{array}\right\}$ from the tower.

It seems to me that sentences like 60 and 61 are partly comments on the appropriateness of the words *chase* and *escape* for the situations being described. These are sentences that would most naturally be used in contexts in which the word *chase* or *escape* had just been uttered.

(60) I didn't 'chase' the thief; as it happened, he couldn't get his car started
(61) I didn't 'escape' from the prison; they released me.

It is important to realize that the difference between assertion and presupposition is a difference that is not merely to be found in the typical predicate words known as verbs and adjectives. The difference is found in predicatively used nouns as well. In the best-known meaning of *bachelor*, for example, the negation-test reveals that only the property of 'having never been married' is part of the meaning proper. Uses of this word (as predicate) *presuppose* that the entities being described are human, male and adult. We know that this is so because sentence 62

(62) That person is not a bachelor

is only used as a claim that the person is or has been married, never as a claim that the person is a female or a child. That is, it is simply not appropriate to use 62, or its non-negative counterpart, when speaking of anyone who is not a human, male adult.

7. Evaluative and Orientative Features

Jerrold J. Katz (1964) and Manfred Bierwisch (1967) have proposed semantic features which treat certain relations between objects and the human beings that deal with these objects. Katz has treated the semantic properties of words that relate to the ways in which objects and events are evaluated, and Bierwisch has proposed ways of associating with words information concerning the ways in which the objects they describe are related to spatial aspects of the language-user's world.

For certain nouns the evaluative information is determinable in a fairly straightforward way from their definitions. This is most clearly true in the case of agentive and instrumental nouns. Many definitions of nouns contain a component which expresses a typical function of the entity the noun can refer to. Thus the lexical entry for *pilot* will contain an expression something like 63

(63) profession: A of [V O A]
where V is 'navigate', with the presupposition that O is an air vessel

and the lexical entry for *knife* will contain such an expression as 64

(64) use: I of [V O I A]
where V is 'cut', with the presupposition that O is a physical object.

For such nouns I assume that the evaluative feature can be automatically specified from the function-identifying part of a definition. A noun which refers to a 'typical' (e.g., 'professional') Agent in an activity is evaluated according to whether the Agent conducts this activity skillfully; a noun which names a typical Instrument in an activity is evaluated according to whether the thing permits the activity to be performed easily. In these ways we can make intelligible our ability to understand expressions like *a good pilot, a good pianist, a good liar, a good knife, a good pencil, a good lock*, etc.

For nouns whose definitions do not identify them as typical Agents or Instruments, the evaluative feature apparently needs to be specified separately. Thus *food* is probably in part defined as 65

(65) function: O of [V O A]
where V is 'eat'.

Food is evaluated according to properties (namely nutrition and palatability) which are not immediately derivable from the definition of *food*, and this fact apparently needs to be stated separately for this item. To call something *a good photograph* is to evaluate it in terms of its clarity or its ability to elicit positive esthetic responses in the viewer, but neither of these notions can be directly derived from the definition of *photograph*. Here, too, the evaluative feature needs to be stated independently of the definition of the word.

The question a lexicographer must face is whether these matters have to do with what one knows, as a speaker of a language, about the words in that language, or what one knows, as a member of a culture, about the objects, beliefs and practices of that culture. Do we know about books that they are used in our culture to reveal information or elicit certain kinds of esthetic appreciation, or do we know about *the word book* that it contains evaluative features that allow us to interpret the phrase *a good book*? Do we understand the expression *good water* (as water that is safe for drinking) because its semantic description has set aside that one use of water as the use in terms of which water is to be generally evaluated, or because we know that for most purposes (e.g., watering the grass, bathing) any kind of water will do, but for drinking purposes some water is acceptable and some is not? These are serious questions, but we can of course avoid facing them by making, with the typical lexicographer, the decision not to insist on a strict separation between a dictionary and an encyclopedia.[15]

The distinction between lexical information about words and non-lexical information about things must come up in dealing with Bierwischian features, too. Let us examine some of the ways in which users of English speak of the horizontal dimensions of pieces of furniture. If we consider a sofa, a table, and a chest of drawers, we note first of all that a sofa or a chest of drawers has one vertical face that can be appropriately called its *front*, but the table does not. For a non-vertically-oriented oblong object that does not have a natural

15. As evidence for the linguistic validity of 'evaluative features' I would accept a pair of words which differ *only* in the evaluative features associated with them. If, for example, English *food* and *feed* could always refer to the same objects but served in the expressions *good food* and *good feed* to refer to food that was *palatable* and *nutritious* respectively, such a pair would provide a good argument for the existence of evaluative features as an aspect of linguistic competence.

front, its shorter dimension is spoken of as its *width*, the longer dimension as its *length*. For the two items that do have a front, the dimension along that front is the *width* (even though it may be the longer of the two dimensions), the dimension perpendicular to the front is its *depth*.

Objects with fronts, furthermore, are typically conceived of as confronted from the outside, as is the case with the chest of drawers, or as viewed from the inside, as with the sofa. The terms *left* and *right* are used according to this inner or outer orientation. Thus the left drawer of a chest of drawers is what would be to our *left* as we faced it, the left arm of a sofa is what would be to our *right* as we face it.

This information is clearly related to facts about the objects themselves and the ways in which they are treated in our culture, and cannot be something that needs to be stated as lexically specific information about the nouns that name them. It seems to me, therefore, that the truly lexical information suggested by these examples is the information that must be assigned to the words *left, right, wide, long* and *deep* (and their derivatives), and that the facts just reviewed about the items of furniture are facts about how these objects are treated by members of our culture and are therefore proper to an encyclopedia rather than a dictionary. It is difficult to imagine a new word being invented which refers to sofas but which fails to recognize one of its facets as its 'front'; and it is likely that if a new item of furniture gets invented, the properties we have been discussing will not be arbitrarily assigned to the noun which provides the name for these objects, but rather the words *wide, left*, etc., will be used in accordance with the ways in which people orient themselves to these objects when they use them. That the orientation is a property of the position- and dimension-words is further demonstrated by the fact that the uses I have suggested are not by any means obligatory. If a 3' by 6' table is placed in such a way that one of its 6' sides is inaccessible, with people sitting at and using the opposite side, the table can surely be described as 6' wide and 3' deep. On the other hand, a sofa that is 2 miles 'wide' would probably impress us more as a physical object than as a sofa and would most likely be described as being 2 miles long.

The phenomena I have been mentioning are to be stated as part of the presuppositional components of the lexical entries for the words *left, wide*, etc. Uses of the word *wide* presuppose that the object being referred to has at least one (typically) horizontal dimension; and that the dimension which this word is used to quantify or describe is either the main left-to-right extent of the object as human beings conceive their orientation to it, if that is fixed, or it is the shorter of two horizontal dimensions. The adjectives *tall* and *short* (in one sense) presuppose as *high* and *low* do not, that the object spoken of is vertically oriented and is typically in contact with (or is a projection out of) the ground. Similarly the noun *post*, as opposed to *pole*, presupposes that the object in question is (or is at least intended to be) vertically oriented and in contact with the ground. Many of the features of spatial orientation treated by Bierwisch will take their place, in other words, in the presuppositional components of the semantic descriptions of words usable as predicates.

There are, however, some spatial-orientation features that appear to enter rather basically into the definitions of nouns. Of particular interest are nouns that identify conceptually *n*-dimensional entities where these are physically realized by *m*-dimensional objects, where $m>n$. Thus a *line* is conceptually one-dimensional, and a *stripe* is conceptually two-dimensional. If a straight mark on a piece of paper is viewed as a *line*, the dimension perpendicular to its length is its *thickness*, but if it is viewed as a *stripe* the second dimension is its *width*. If the stripe has a third-dimensional aspect (e.g., if it is drawn with heavy paint), it is that which one speaks of as its *thickness*. These are matters that seem to be related to the 'meaning' of these nouns rather than to presuppositions about the objects they name.

8. Functional Shift

Syntactically and semantically different uses of the same word type should be registered in the same lexical entry whenever their differences can be seen as reflecting a general pattern in the lexical structure of the language. I shall call attention to certain situations in which a word that is basically a noun can also be used verbally, and a situation in which a verb of one type is used as a verb of another type.

I have already suggested that the 'sentential' portions of the definitions of agent and instrument nouns serve to provide the evaluative features associated with these nouns. In many cases they also serve in identifying the verbal use of these same nouns. If, for example, the word *pilot* is defined in part as one who flies an airplane, a dictionary entry must show some way of relating this aspect of the meaning of the noun to the meaning of the associated verb. Perhaps, for example, in connection with the activity characterized as 66

(66) V O A
where V = navigate; presupposition:
O = air vessel

one might wish to state that the noun *pilot* is the name given to one who professionally serves the Agent role in this activity, the verb *pilot* has the meaning of the verb in this event-description. If the word is further represented as basically a Noun, a general understanding of 'metaphor' will suffice to explain why the verb *pilot* can be used to refer to activities that are 'similar to', not necessarily identical with, the activity of a pilot in flying an air plane.

If the noun *lock* is defined as a device which one uses to make something unopenable, then that is related to the fact that the *verb lock* means to use such a device for such a purpose. If *plaster* is defined as something which one attaches to a surface for a particular range of purposes, then that fact should be shown to be related to uses of the *verb plaster.* I have no proposal on how this is to be done; I merely suggest that when both the verbal and the nominal use of a word refer to events of the same type, the event-description should, other things being equal, appear only once in the lexicon.

Certain fairly interesting instances of verbal polysemy seem to have developed in the following way. Where one kind of activity is a possible way of carrying out another kind of activity, the verb which identifies the former activity has superimposed onto it certain syntactic and semantic properties of the verb which identifies the second or completing activity.

Thus the verb *tie* refers to particular kinds of manipulations of string-like objects, with certain results. In this basic meaning of the verb, it occurs appropriately in sentences 67 and 68

(67) He tied his shoestrings
(68) He tied the knot.

The act of tying things can lead to fastening things, and so an extension of the verb *tie* to uses proper to a verb like *fasten* or *secure* has occurred. The verb can now mean to fasten something by performing tying acts, and it is this which accounts for the acceptability of *tie* in 69.

(69) He tied his shoes.

Shoes are simply not in themselves the kinds of objects that one manipulates when tying knots.

In this second use the verb *tie* continues to describe the original activity, but it has been extended to take in the result of such activity. The feature that characterizes this second use, then, will be something like 70.

(70) extension: Result (replace *fasten*).

The verb *smear*, to take another example, refers to the activity of applying some near-liquid substance to the surface of some physical object. The activity of smearing something onto a thing can have the result of covering that thing. The word *smear* has, in fact, been extended to take on the syntax and semantics of *cover*. Thus the 'original' and the extended uses of *smear* are exemplified in the following sentences:

(71) He smeared mud on the fender
(72) He smeared the fender with mud.

The difference by which 71 and 72 are not quite paraphrases of each other is found in the meaning of sentence 73

(73) He covered the fender with mud.

By claiming that the second use of *smear* is one in which the properties of *cover* have been superimposed we have accounted for the addition to 72 of the meaning of 73, and simultaneously we have accounted for the fact that the extended use of *smear* takes (as does *cover*) the Goal rather than the Object as its direct object, setting the latter aside in a preposition-phrase with *with.*[16]

The verb *load*, let us say, means to transfer objects onto or into a container of some sort. The activity of loading can lead to the filling of that container, and so the verb *load* has taken on the additional syntactic and semantic functions of *fill*. In this way we can account for the use of *load* in sentences 74 and 75 and the similarities between 75 and 76

(74) He loaded bricks onto the truck
(75) He loaded the truck with bricks
(76) He filled the truck with bricks.

The verbs *smear* and *load* have the same co-occurrences in their extended meanings as in their non-extended meanings. The verb *tie* is not like that: in its *fasten*-extension it takes nouns that are not appropriate to its original

16. I believe that 72 can be read as a paraphrase of 71. If this is so, a correct description of the situation must be that in its original meanings, *smear* permits the direct-objects choices seen in either 71 or 72, but in its extended meaning it permits only that exemplified by 72.

sense. This means that the description of the feature indicated in the extended use will have to be interpreted (by lexico-semantic rules) in such a way that when the two verbs fail to take the same cases, those of the verb which identifies the resulting action are dominant, the characteristics of the event described by the other verb taking their place among the *presuppositions* of the verb in its extended sense. Thus it is presupposed of the *fasten*-extension of *tie* that the Object is something which can be fastened by an act of tying.

If the type of extension that I have been discussing in this section can be shown to be a quite general phenomenon of lexical systems (at present it is little more than a suggestion, the 'evidence' for its correctness being rather hard to come by), then perhaps we can use this concept to eliminate certain problems connected with what Gruber has called 'incorporation'. *Leap* is a verb which takes, in Gruber's system, a phrase in *over*, as seen in 77 and 78

(77) He leaped over the fence
(78) He leaped over the line.

The preposition *over* can be 'incorporated" into *leap*, but only in the understanding that the associated noun is an obstacle[17]; thus, the preposition *over* may be absent from 77, but not from 78. The theoretical issue here has to do with the way in which the process of preposition-incorporation is to be sensitive to the size relationship between the entities identified by the subject and prepositional-object nounphrases. My interpretation is that there is an *overcome*-extension to the word *leap*, and I claim that this accounts simultaneously for the 'obstacle' presupposition and for the non-occurrence of *over* in the extended-sense sentence.[18]

9. Deep-Structure Acceptability

Facts about lexical items that relate to the formal properties of sentences can be separated into two sets: requirements on the deep-structure and requirements on the surface structure. The former determine the acceptability of a given word in deep-structures of certain types; the latter specify those grammatical modifications of sentences whose operation is determined by lexical

17. Gruber (1965, p. 24). The condition for incorporation of *over* into *leap*, in Gruber's words, is this: "The object of the preposition must be of significant height with respect to the subject".

18. This argument is certainly not directed against Gruber's incorporation process as such, only against the proposed need to state separately its applicability to words like *leap, jump, hop,* etc. The quite literal 'incorporation' of *over* in *overcome* has not escaped my notice.

information. The surface conditions are provided in the grammar in the form of the rules which convert deep structures into surface structures (transformational rules), and possibly, in some cases, by the elaboration of special constraints on surface structure.

I shall take the position that content words may all be inserted as predicates, and that their realization as nouns, verbs or adjectives is a matter of the application of rules. Therefore we need not consider part-of-speech classification among the types of information relevant to the lexical insertion into deep structures. What Chomsky (1965, pp. 95–100) has referred to as 'strict subcategorization' corresponds to what I have treated here in terms of the number of arguments a predicate takes and their case structure. What Chomsky (1965, pp. 148ff) has referred to as 'selection' is described here with the concept presupposition and is taken as being more relevant to semantic interpretation than to lexical insertion.

The deep-structure requirements that are of chief interest for this discussion, then, are those of the type Perlmutter has been referring to as 'deep structure constraints' or 'input conditions'.[19]

Examples, due to Perlmutter, are the requirement for *budge* that it occur in a negative sentence, as shown in grammaticality judgments on 79 and 80;

(79) *I budged
(80) I didn't budge

the requirement for *lurk* (discussed earlier) that its Agent be non-coreferential to the Agent of the 'next higher' locutionary verb; or for *try* that its Agent be coreferential to the (eventually deleted) subject of the 'next lower' sentence, as suggested by 81 and 82

(81) *I tried [for you] to find it
(82) I tried [for me] to find it.

I have included deep-structure constraints in this survey of types of lexical information, but I have nothing new to say about them. I would like to suggest, however, that it may not be necessary to require the extent of detail which Perlmutter envisions or the transformational apparatus which that some-

19. My knowledge of Perlmutter's work on deep-structure and surface-structure constraints comes from Perlmutter's presentations at the January, 1968, San Diego Syntax Conference and from references in papers and presentations by J. R. Ross.

times entails. Where Perlmutter requires that the Agent of *try* match the *Agent* of the embedded sentence, it may only be necessary to require that the coreferential nounphrase in the embedded sentence be the one which is destined to be the subject of that sentence. And where Perlmutter requires sentence 83 to be derived transformationally from the structure underlying 84,

(83) He tried to be misunderstood
(84) He tried to get to be misunderstood

this may not be necessary if *try* is merely described as a verb which expresses, of its Agent subject, the intension and attempt to bring about the situation identified by the embedded sentence. This may be necessary in order to account for the way in which we understand sentence 85,

(85) He tried to seem cheerful

a sentence which cannot be straightforwardly paraphrased in such a way as to reveal an underlying agentive notion in the embedded sentence.

10. Government

Once a specific predicate word is inserted into a deep structure, its presence may call for certain modifications in the rest of the sentence. The typical case of this is what is known as 'government'. For English the operation of the rules establishing 'government' associates prepositions with noun-phrases and 'complementizers' with embedded sentences and their parts.[20]

Thus–to consider only the association of prepositions with noun-phrases –we speak of *giving something to somebody, accusing somebody of something, blaming something on somebody, interesting somebody in something, acquainting somebody with something,* and so on. It is certain, of course, that many of the facts about particular choices of prepositions and complementizers are redundantly specified by other independently motivated features of predicates or are determined from the nature of the underlying case relationship, so that a minimally redundant dictionary will not need to indicate anything about the form of governed constituents directly. Until it is clear just what the needed generalizations are, however, I propose using the brute-force method

20. The term 'complementizer' is taken from Peter Rosenbaum (1968), and refers to the provision of *that, -ing,* etc., in clauses embedded to predicates.

and specifying the prepositions one at a time for each verb and each case relationship.

11. Transformationally Introduced Predicators

Certain lexical items can be used as predicates semantically, but cannot themselves occur as surface predicate words. Such words will appear in the syntactic position expected of some other constituent (usually, I think, that of the Result constituent), and there must therefore be lexically associated with them some predicator word, a word that is capable of bearing the tense and aspect properties that can only be attached to verb-like elements. The constructions I have in mind are those of the type *have faith in*, *give credence to*, *be loyal to*, etc., and the predicator words I have in mind include *be*, *do*, *give*, *have*, *make*, *take* and a few others.

12. Subject and Object

One of the subtlest but at the same time most apparent aspects of the syntax of the predicate is the set of ways in which the subject/verb and verb/object constructions can be set up. These constructions have as their purpose–generally, but not always–a focusing on one terminal of a multi-argument expression, a focusing on one particular role in the situation under discussion.

I have suggested elsewhere (see Fillmore 1968, note 6) that the normal choice of subject is determined by a hierarchy of case types, and only those predicates which require something 'unexpected' as their subjects need to have such information registered in their lexical entry. I believe that, at least in the vast majority of cases, those noun-phrases capable of becoming direct objects of a given verb are at the same time the ones which can appear as the subjects of passive constructions made with the same verb. Thus such information at least does not need to be stated twice, and at best may not need to be stated at all for most verbs, since it is most likely subject to rather general statements about combinations of cases.[21]

There is an extremely interesting set of subject- and object-selection facts that seem to operate in connection with expressions of 'quantity' and 'contents'. The verb *sleep* refers to an 'activity' of an animate being in a particular place, where the one who sleeps is typically mentioned as the subject; but

21. Incidentally, verbs which are obligatorily passive are verbs with one associated nounphrase designated as 'direct object' but none as 'normal subject'. An automatic consequence of this situation is that the expression will be cast in the passive form.

when the focus is on the place and at issue is the *number* of different beings that can sleep in that place, the verb permits the Place nounphrase to appear as subject. We see this in sentence 86

(86) This houseboat sleeps eight adults or sixteen children.

The verb *feed* has its use as a 'causative' of *eat*, but it can be used to express other kinds of relations with eating, too. To indicate the typical relation between the Agent and a description of the 'contents' of his eating activity, we use sentences like 87

(87) The child feeds on raisins.

There is a relation between a recipe (which identifies, among other things, quantities of food) and the number of people who can (with satisfaction) eat the food one prepares by following the recipe, and this is the relation we see expressed in a sentence like 88

(88) This recipe feeds eight adults or four children.

In the case of *feed*, the connection between the Place and a quantity of 'eaters' requires the latter quantity to be a 'rate'. Thus we get sentences like 89

(89) This restaurant feeds four hundred people a day.

It is not clear to me how one can capture facts of the sort I have just suggested in any way short of providing (in effect) separate lexical descriptions for each of these uses. It is clear, anyway, that examples 86–89 cannot be understood as merely exemplifying 'causative' extensions of these verbs.

13. Sample Lexical Entries

In this last section I exhibit extremely tentative suggestions for the lexical entries for *blame* (in two senses[22]), *accuse*, and *criticize* (in one sense). The feature-pair ± Locutionary indicates whether the verb does or does not refer to a linguistic act; the feature + Performative marks verbs which can be used in first-person simple-present utterances in the performance of an explicitly

22. There is a third use of *blame* in which it refers to a linguistic or otherwise symbolic act. Thus, if John wrote an offensive letter himself and did something which gave others to believe that I has done it, I could report that John blamed the letter on me.

marked illocutionary act[23]; the feature-pair ± Momentary indicates whether or not the verb can be used, in the affirmative, to refer to an event that takes place at a specific point in time. The variables *x, y,* and *z* are argument variables whose case-role and preposition-selection properties are indicated by first, second or third position in the next two items. All are normal transitive verbs, and the normal subject-selection for active sentences is in each case the noun-phrase that fills the Source role. For *blame* the direct object is either the Goal or the Object, since we have both '*blame y for z*' and '*blame z on y*'; for *accuse* the direct object is necessarily the Goal noun-phrase; for *criticize* it is the Goal noun-phrase if that is explicit in the sentence, otherwise the Object noun-phrase. The verbs *accuse* and *criticize* require the Goal noun-phrases to identify human beings; *blame* does not, since one can blame an event on, e.g., an inanimate force. The items identified as 'zero for indefinite' and 'zero for definite' indicate the conditions under which the explicit mention of an argument may be omitted, in the sense mentioned in Section 5.

There are many types of lexical information not found in these sample entries, for example those mentioned in sections 7 and 8 above. There are many facts about these particular verbs that are ignored in these descriptions provided here, for example the information that the situation named by the Object noun-phrase for *blame* and *criticize* is necessarily understood as something for which factuality is claimed (notice 90 and 91); that the linguistic event capable of constituting a criticism is necessarily more complicated that the linguistic or gestural event capable of constituting an accusation; and that the presupposed seriousness of the offense for an act of criticism is not as great as that appropriate for acts of accusation (compare 92 and 93). It is hoped, nevertheless, that from these examples some suggestions can be gleaned for the design of canonical representations in a lexicon.

(90) I accused John, who has never left Ohio, of taking part in the Berkeley riots.

(91) *I criticized John, who has never left Ohio, for taking part in the Berkeley riots.

(92) I accused John of murdering his mother-in-law.

(93) I criticized John for murdering his mother-in-law.

---

23. In the sense of Austin (1965).

BLAME (i): –Locutionary, –Momentary

arguments:	*x, y, z*
cases:	Source & Experiencer, Goal, Object
prepositions:	by, on, for
normal subject:	*x* direct
object:	*y* or *z*
presuppositions:	*x* is human
	z is an activity *z'* or the result of *z'*
	x judges [*z'* is 'bad']
meaning:	*x* judges [*y* caused *z'*]
zero for indefinite:	*x*
zero for definite:	*z*

BLAME (ii): –Locutionary, –Momentary

arguments:	*x, y, z*
cases:	Source & Experiencer, Goal, Object
prepositions:	by, on, for
normal subject:	*x*
direct object:	*y*
presuppositions:	*x* is human
	z is an activity
	z' or the result of *z'*
	x judges [*y* caused *z'*]
meaning:	*x* judges [*z'* is 'bad']
zero for indefinite:	*x*
zero for definite:	*z*

ACCUSE: +Performative, +Locutionary, +Momentary

arguments:	*x, y, z*
cases:	Source & Agent, Goal, Object
prepositions:	by, Ø, of
normal subject:	*x*
direct object:	*y*
presuppositions:	*x* and *y* are human
	z is an activity
	x judges [*z* is 'bad']
meaning:	*x* indicates [*y* caused *z*]
zero for indefinite:	*x*
zero for definite:	*z*

CRITICIZE: +Locutionary, +Momentary

arguments:	*x, y, z*
cases:	Source & Agent, Goal, Object
prepositions:	by, Ø, for
normal subject:	*x*
direct object:	*z* if *y* = Ø, otherwise *y*
presuppositions:	*x* and *y* are human
	z is an activity *z'* or the result of *z'*
	x judges [*y* caused *z'*]
meaning:	*x* indicates [*z'* is bad]
zero for indefinite:	*x* and *z*; and y if *z* is a 'result' and $z \neq 0$
zero for definite:	none[24]

24. Sponsored in part by the National Science Foundation through Grant GN-534.1 from the Office of Science Information Service to the Computer and Information Science Research Center, The Ohio State University. This article has also appeared in: *Semantics: An Interdisciplinary Reader in Philosophy, Linguistics, Anthropology and Psychology* (ed. by D. D. Steinberg and L.A. Jakobovits), University of Illinois Press Urbana (Ill.), 1969.

5

The Case for Case Reopened

This paper responds to a number of comments and criticisms of my writings on case theory, clarifying what seemed to me to be misunderstandings, and presenting a few new arguments as well as a few adjustments in the theory.

It begins with a summary of what I considered the position of case theory within theories of semantics and grammar. As a contribution to semantics, case theory is described as concerned with internal *as opposed to* external *semantics, where internal semantics tries to characterize the information that is being communicated, in opposition to external semantics, which is the relation of sentences to matters of truth and entailment. Within internal semantics, the emphasis is syntagmatic rather than paradigmatic, having to do with the functional relations between parts of a clause rather than with relations between linguistic expressions and their alternatives (synonymy, antonymy, hyponymy, etc.).*

I had declared "the centrality of syntax" as an underlying principle of case-grammatical thinking, by which I meant that when it came to the constituents of a clause, the description should proceed from their function to the morphology, rather than the other way around. The intended context had to do with descriptions of morphological categories and (e.g., the dative case) and their uses. This was taken by some as contradicting what I had said elsewhere about the centrality of semantics.

Some authors felt that the kinds of arguments I had given for justifying a contrast between case roles was merely a function of the English lexicon, where the same word form could have a wide variety of combinatory possibilities, possibilities that would be built around separate words in many other languages. I point out the arguments I had been using were for showing how such structures could be discovered even with a language like English, but that in languages that used different word stems or perhaps causative or applicative morphology for what in English can be expressed with homonymous forms, the explanation of those differences would be equally appropriate for pointing out case role differences.

Originally published in 1977

On two important issues I surrendered to my critics: first, it was wrong for me to speak in the earliest writings of an animacy contrast as being relevant to the definition of deep cases (as with Agent and Experiencer as opposed to Instrument and Patient); and second, I had to recognize that there is an aspect of sentence interpretation that does indeed make use of the two grammatical functions subject *and* object, *and that clause types that differed in respect to the realization of particular cases in these functions had semantic differences clearly related to such distinctions. I tried to account for these effects by a system of speaker decisions on what we might call* rhetorical structure*–something I would not do now–involving the assignment of salience to some constituents rather than others and appealing to certain generalizations, expressed as a case hierarchy, for grammatically expressing the elements given salience.*

Central to the reasoning was an account of the verbs of commercial acts, important in that in events named by these verbs there are always two active participants (hence two agents?) and two moving entities (hence two themes?). Briefly, the argument is that while our understanding of a commercial event involves two active participants, the verbs we can choose select one or the other of them (the buyer *or the* seller*) as bearing a point of view, or perspective, expressed in the nucleus of the sentence, and they can select one or the other of the thematic entities (*goods *or* money*) as being the focused-on theme in the transaction. Thus* buy *highlights the buyer and the goods,* sell *highlights the seller and the goods,* pay *highlights the buyer and the money, and so on. There are of course other case relationships that these elements simultaneously bear: since the seller is the source of the goods and the goal of the money, and the buyer is the source of the money and the goal of the goods, once Agent and Theme are chosen, the other (optional) elements have to be expressed with the relevant prepositions. Thus, you* pay *the money* to *the seller (since it's the seller that receives the money), you* buy *the goods* from *the seller (since it's the seller that surrenders the goods), and so on; and there has to be a separate preposition for the secondary theme, namely* for.

1.

Several years ago I wrote a fairly long article (Fillmore 1968) with the punning title, "The Case for Case." In it I presented a notion I referred to as *deep structure cases,* together with a number of proposals on how the notion could be incorporated into a generative grammar. Since that time, I have become aware–through the many published and unpublished reactions to the piece that have reached my desk–not only of some of my own mistakes, but also of a number of ways in which readers have misunderstood my intentions or have acquired expectations for the theory which went beyond anything I had in mind.

Here is what I am going to try to do in this chapter: I will locate the concept of deep cases within traditions of semantic and grammatical inquiry; I will review a few of the basic assumptions of case theory and will point out what I originally thought of as its advantages; I will discuss some of the most important challenges to the theory, including the serious one of how one can know what the cases are and how many of them there are; and I will fail to offer a satisfying solution to this problem. I will instead propose a new interpretation of the role of cases in a theory of grammar and a new method of investigating the question of their number and identity. I consider this new interpretation as a position in the theory of grammatical relations and as a position in semantic theory with which one could associate this slogan: *Meanings are relativized to scenes.*

To illustrate my points, I will mainly use the same examples that I have used in earlier writings, including, of course, many that I borrowed from other authors. This is not because I can't think of any new ones, but because I wish to show with examples that might be familiar some of the ways in which the new interpretation is different.

2.

Within semantics in general, the notion of deep cases is a part of what might be called *internal*, as opposed to *external*, semantics; that is, it concerns, not the semantics of truth or entailment or illocutionary force, but rather the semantic nature of the inner structure of a clause. Within internal semantics, the concern is *syntagmatic* rather than *paradigmatic*; that is, deep cases are among the types of semantic relations that elements of sentence structures have with each other in context, rather than with the system of contrasts and oppositions that differentiate constituents paradigmatically. The concern is with the inner structure of clauses rather than with the semantics of interclausal connections through the devices of coordination and subordination.

Within grammatical theory the concept of deep cases can be thought of as a contribution to the theory of grammatical levels, to the theory of grammatical relations, to the description of valences and collocations, and to the general theory of the functions of sentence constituents. The proposal is that there is a level of structural organization for a sentence which is distinct from what is usually thought of as a semantic representation and which is distinct likewise from the familiar notions of deep and surface structure syntactic representations. The theory is concerned with the determination of the nuclear grammatical relations in a sentence–subject, object, and indirect object–in that it addresses itself to the question of how particular aspects of the meaning of an utterance determine which constituent will appear as the (deep structure) subject, which as the object. The theory of cases can also be seen as offering at least part of the *semantic valence* descriptions of verbs and adjectives, comparable to the syntactic valence descriptions given by certain European linguists (Tesnière 1959, Helbig 1971, Helbig and Schenkel 1969, Emons 1974). And, lastly, it may add something to the theory of the functions of sentence constituents. Katz (1972, p. 113) has distinguished three kinds of sentence-constituent functions: the *grammatical*, illustrated by the notions of 'subject' and 'object'; the *rhetorical*, illustrated by such oppositions as 'given' versus 'new', 'topic' versus 'Comment', etc.; and the *semantic*, illustrated by such notions as 'agent', 'recipient', 'means', 'result', etc. In his discussion of these functions, Katz claims that Chomsky's failure in the Extended Standard Theory (Chomsky 1970) is one of confusing the grammatical and the rhetorical functions, and that my failure in "The Case for Case" is that of confusing the grammatical and the semantic functions. There is, I believe, a fourth way of looking at the functional structure of the parts of sentences, something for which such words as *orientation* and *perspective* suggest what I have in mind. The parts of a message can be divided into those that are "in perspective" and those that are "out of perspective". My current position is that it is the orientational or perspectival structuring of a message which provides the subject matter for the theory of cases, and that the case notion figures very differently in grammatical description from what I originally had in mind.

One essential feature of the theory of deep cases is the *case frame* (Fillmore 1968, p. 27), the function of which is to provide a bridge between descriptions of situations and underlying syntactic representations. It accomplishes this by assigning semantico-syntactic roles to particular participants in the (real or imagined) situation represented by the sentence. This assignment determines or constrains the assignment of a *perspective* on the situation by means of what I have called *subject selection principles* and the *case hierarchy*.

Some of the Subject Selection principles seem to be language-universal. Thus, given certain qualifications for the interpretation of ergative systems, one candidate for a universal Subject Selection Principle is this: If there is an agent which is brought into perspective, the nominal which represents it must be the (deep) subject.

Other Subject Selection principles appear to be language-specific. Japanese, according to Kuno (1973, p. 31), and German, according to Rohdenburg

(1970), do not allow certain kinds of enabling or occasioning causes to be chosen as subjects of their sentences, whereas English does tolerate such choices, as in sentences like those in 1:

(1) a. Fifty dollars will buy you a second-hand car.
b. The smell sickened me.
c. The accident killed the woman.

Still other Subject Selection principles appear to be word-specific. This appears to be true, for example, of (at least) one member of the pair *regard* and *strike*, as in Chomsky's examples (1965, p. 162) given here in 2:

(2) a. I regard John as pompous.
b. John strikes me as pompous.

Lastly, but not surprisingly, Subject Selection principles are capable of changing in time. Jespersen (1924, p. 160) speaks of the changes in the history of English from expressions of the type seen in 3a and 3b to expressions of the type seen in 3c and 3d as involving changes in the *meanings* of these verbs:

(3) a. Me dreamed a strange dream.
b. Me like oysters.
c. I dreamed a strange dream.
d. I like oysters.

There is, of course, a way of construing the word *meaning* so that Jespersen's description can seem sensible; but one could more revealingly say that what changed were the workings of the Subject Selection principles operating in the language, these changes having to do with the interaction of ordering principles and case-marking principles in a period after which most surface case distinctions on nouns and pronouns were lost.

The advantages that I saw in the notion of deep-structure case were that case-structure descriptions of words and sentences offered a level of linguistic organization at which universal properties of lexical structure and clause organization were to be found, and, moreover, that such descriptions were in some sense intuitively relatable to the ways people thought about the experiences and events that they were able to express in the sentences of their language.

3.

It was misleading for me to use the phrase *case grammar* to describe the proposals I made in "The Case for Case." My proposals did not cohere into a model of grammar. Instead, they were suggestions about a level of organization of a clause that was relevant to both its meaning and its grammatical structure; that provided a way of describing certain aspects of lexical structure; and that offered convenient classifications of clause types. From the fact that I have frequently received mail asking me questions like, "How does case grammar handle intonation?", I have become aware that my writings somehow gave the impression that case grammar so-called was being presented as a general model of linguistic structure. Nowadays I try to be more careful about the phrase 'case grammar'.

Criticism of the deep case proposals have come from many places and represent many points of view. Some of them are based on misunderstandings (possibly invited misunderstandings); some of them are valid but require "patching" rather than deep changes in my position; and a number of them are very much worth worrying about.

When I wrote "The Case for Case" the Generative Semantics position had not yet been formulated, or, rather, had not yet been formulated in my hearing except in its preliminary version as Abstract Syntax. In the section of my article which touched on surface case morphology, I described the traditional way of treating cases, which consisted in identifying the cases through their forms and then, one by one, describing their functions within larger constructions. What I suggested was that things should be turned around, so that the organization of the sentence as a whole should be taken as the framework within which the functions of the individual grammatical morphemes could be stated. As a catchword for this position I spoke of the *centrality of syntax*. Walter Cook, in his article "A Set of Postulates for Case Grammar" (1972), claims to be taking a position in opposition to mine in a postulate which assigns centrality to semantics, not syntax. I suspect that if I knew precisely what it meant, Cook's is a position I might take, too. But, in any event, it was not an opposition between semantics-at-the-bottom and syntax-at-the-bottom that I had in mind, but rather between analysis that begins with the morpheme and analysis that begins with the sentence.

I have used various forms of arguments in trying to justify specific claims about deep case distinctions. One of these argument forms has two steps. In the first step we recognize an ambiguous sentence whose ambiguity can be accounted for only by assuming that one of its nominals is interpretable as bearing either of two semantic roles in the sentence. As an illustration of this point, we can consider the ambiguous sentence 4a. To see that this sentence is ambiguous, imagine a situation in which you see that I have a letter on my desk, that you see me make a copy of it, and that then you hear me say sentence 4b. If you speak the same version of English that I do, you can't know for sure, from my instructions, which of the two letters I have in mind.

(4) a. I copied the letter.
b. Point to the letter which I copied.
c. I copied this from that.

The second step in this argument form consists in exhibiting the same verb in a sentence where it takes two different nominals, each with just one of the two semantic roles sensed in the earlier ambiguous sentence. This property is found in our verb *copy*, as can be seen in sentence 4c, in which exactly the two previously sensed roles noticed in 4a are parcelled out among two separate nominals. Having both types of evidence–as provided in the two steps of this argument–makes it believable that we are dealing with two different case roles rather than with simply a matter of vagueness.

In a second kind of argument, also with two steps, I have suggested that we may be dealing with different case relationships whenever we find a single verb collocating with two seemingly disparate classes of nominals in a given grammatical relation, with the nominals from the two classes seeming to exemplify different semantic roles in their sentences. Examples for this point are the possibilities for the subject-nominal in sentences 5a and 5b. And, again, the argument can be taken as completed if, as step two, we can find a single sentence with the same two roles parcelled out between two separate nominals, as in 5c.

(5) a. My foot hurts.
b. This shoe hurts.
c. This shoe hurts my foot.

These kinds of arguments have seemed particularly weak to a number of European critics, who have suggested that the method of analysis is based on accidental properties of English words. Other languages, this particular counter-argument goes, might use different words for the different senses of *hurt* and *copy*, and so, for purely nonsystematic reasons, the method would yield different results for these other languages.

This objection, too, I take as a misunderstanding. These arguments are not offered as definitions of cases, but rather as steps for pointing out case distinctions in sentence sets where sense differences are only, or are primarily, matched by case role differences. I have been just as interested in similarities in underlying case structure in sentences with different verbs and different relational organization. In a third kind of argument that I have used, different surface verbs are taken from single vocabulary fields–pairs like *rob* and *steal* or *buy* and *sell*–verbs which have matching (or partly matching) case structures but different assignments of grammatical relations. I feel sure, in other words, that an analyst looking at, say, Swedish, would be able to discover, in expressions of physical pain, roles involving such things as the source of the pain, the location of the pain, the experiencer of the pain, and so on; and this could be done independently of whether the same verb or different verbs were used, independently of whether the same or different choices were made in the selection and orientation of the cases.

A criticism that has come from some European workers in Valence Theory (Emons 1974 and Panevová 1974) has been that I have obscured the various senses in which it is possible to speak of optional constituents in a sentence. Sometimes some aspect of an event or situation is a part of the speaker's and the hearer's understanding of the meaning of a sentence, yet there is nothing in the sentence which expresses it; and sometimes the absence of a constituent in the surface sentence reflects the absence of the associated notion from the conceptualization that is being communicated. It is true that there have been some places where I was wrong in detail in the description of given sentences in just this regard; but my intention was that all of the various senses of optionality could be accounted for by the fact that the system I proposed had *case frames*, indicating the case notions conceptually present in a sentence; *case frame features*, indicating the case notions that could be combined in construction with a given lexical item; and *deletion transformations*, by which, under various conditions, a given constituent could be, or maybe had to be, absent from the surface structure. I think that in principle my proposals take care of optionality sufficiently well.

Several specific suggestions that I made in "The Case for Case" seem to have been bad ones, and I have been quick to abandon them. I proposed, for example, that all nominals in English sentences are initially provided with prepositions; this position required that the processes of forming subjects and direct objects had to involve preposition deletion. My motivation was partly that of wanting to make English and Japanese seem more comparable than they might have otherwise. Japanese has *postpositions* marking all nominal functions in a sentence, *including the nuclear grammatical relations* (subject and object). I thought that the basic typological differences between the two languages could be captured as follows: In deep structure, English has the verb at the beginning, Japanese, the verb at the end, of each sentence; the language with the verb at the beginning has prepositions on all the nouns; the language with the verb at the end has postpositions on all the nouns. Apart from these differences–which a mirror would show are really similarities–the two languages differ in that the verb-initial language–English–has a Subject-fronting rule, by which the verb ends up in second position, and it has preposition-deletion processes for the subject and direct object. These two language types, then, can be compared with a third type, exemplified by Tagalog, which lacks Subject Fronting and which keeps prepositions in front of all its associated nominals.

As part of an informal general typological sketch, this way of looking at things is not bad, I think; but the specific proposals I had for the assignment of the individual lexical preposition made the system look fairly inefficient. In particular, as I tried to construct a grammar using these proposals, my initial association of the preposition *by* with the agent case had to yield in favor of a more complicated principle associating *by* with the highest-ranking case in the sentence, whatever that case might have been. This decision made it possible to account correctly for the *by* in *eaten by George, destroyed by fire,* and *assumed by everybody* (as long as it was constrained to allow *known to me,* etc.). In the end, however, such provisions did not look any better than an account according to which the preposition *by* gets introduced by means of a Passive transformation.

In "The Case for Case" I described the agent and the dative cases as necessarily animate. This, by a curious kind of bad logic, led some people, for example me, to assume that obligatory animacy in a verbal complement implied that the associated nominal had to be in one of these two cases. By that kind of reasoning, the subjects of *die* and *melt* (to use examples from Huddleston's 1970 critique) were dative and patient (object), respectively, and the two sentences in 6 would have to be given different case structure analyses:

(6) a. The man died.
b. The snow melted.

I am now more careful about keeping relational notions and categorial notions distinct.

Some workers, for example Nilsen (1972), have provided binary opposition analyses of the basic cases, analyzing agent and instrument as animate and inanimate causes, respectively; and experiencer (dative) and patient (object) as animate and inanimate effects, respectively. The initial appeal of this analysis fades when one realizes that it is a confusion of categorial and relational notions, but begins to reappear when one tries to see it as an account of the basic properties of the prototype scenes in terms of which actions and experiences can be structured.

In an attack on case grammar recently published in Germany, Peter Finke (1974) concentrated on what he took to be the categorial implications of cases, and interpreted case theory as a version of many-sorted logic. To say that a verb takes an agent and a patient is to say, Finke would claim, that the two arguments of the verb have to meet different *sortal specifications*, and that in general cases are to be defined in terms of sortal specifications. Agents are things having one set of properties, instruments are things having some other set of properties, and so on.

Finke's main point, if I understand him correctly, is that a case grammar makes an ontological commitment about the number of sorts of objects that can exist in a universe, that number equal to the number of cases. If such a strong claim is implied by the theory, according to Finke, then the fact that case grammarians have not been able to come up with a common and stable collection of cases shows that the theory is not to be taken seriously. The reply is, of course, that even if some universe contained only one sort of object—say, human beings—the role-identifying function of the cases could still be maintained. One person could pick up another person, use that person's body for knocking down a third person, that third person could feel embarrassment, and so on. In a universe with only one sort of object, in short, the case relations of agent, instrument, patient, and experiencer could all be easily imagined. Perhaps it was my misleading statements about animacy that created the impression that case notions were to be understood as categories rather than as types of relations.

A challenge to case grammar that I have recently heard comes from a writer (Raible 1975) who was concerned with how case theory could account for the very complicated set of case functions in a language like Finnish. The author of this challenge had taken my suggestion that when looking at a system of cases we should distinguish case *forms* from case *uses*, and that we should build a theory of deep cases whose repertory of case concepts was more or less equivalent to the repertory of case uses. Finnish presents special problems, and in two ways. First, because of the system of locative and directional cases which combine, in one and the same surface case category, both locational notions and assumptions about the topology of the reference object (the entity in terms of which something is being located or with respect to which something is being oriented); and, second, because there is a semantic difference, in what is thought of as the direct object, between choosing the nominative, the accusative, or the partitive case. The difference between choosing the accusative rather than the partitive, for example, appears to involve such notions as definiteness, completion, and totality. My answer has to be, of course, that the morphemes in surface case systems encode more than nominal functions, and that therefore an account of the uses of surface-structure cases requires more than a theory of deep cases.

4.

There are four more kinds of criticisms of case theory that I wish to mention, two that I do not know how or whether to worry about, and two that seem clearly very serious indeed. The first of the puzzling ones is the collection of criticisms that take the form, "I can do anything you can do"–the argument of *mere notational variance*. One version of the argument goes like this : I have claimed that in the structure underlying expressions containing a verb like English *seem* there is what I call an *experiencer*, the person for whom something seems; and I have claimed that this sometimes is manifested on the surface as a *to*-phrase, as in 7:

(7) a. To me, Harry seems intelligent.
b. It seems to me that Harry is intelligent.

This, the argument goes, is pointless. There is no particular advantage to giving theoretical prominence to terms like experiencer, agent, instrument, and the rest. We could just as well state that *seem* is a *mental state verb*, meaning that it has a semantic marker 'mental'; and we can then say that selection restrictions for verbs containing this marker require one of their coconstituents to be 'animate' nominal; we can say of the verb *seem* that it has a selectional restriction requiring its associated animate argument to be part of a preposition phrase marked by *to*; and we could even provide a principle which declared that, for any argument of a mental verb which is obligatorily animate, a semantic marker 'experiencer' can be added to the representation of that nominal in the finished semantic reading of the sentence.

This argument (found in Chomsky 1970, Katz 1972, and Mellema 1974) has something to say for it, of course. It is a version of the view that a change in a scientific paradigm should be tolerated only when the possibilities of the existing paradigm have been exhausted *and* when a more satisfactory paradigm can be shown to exist. I sympathize with this view in general; my position, however, is that alternative paradigms–or even alternative notations–should be valued for the kinds of questions they force an analyst to ask. A strong assumption about deep structure cases forces the analyst to ask certain questions about the number and variety of the semantic functions of the parts of sentences. Whether the answers to these questions can be *written down* within the terms of some other paradigm does not always seem important to somebody primarily concerned with whether the questions are important ones and whether they can be answered at all.

Next there are the trivializing objections to case grammar leveled by Ray C. Dougherty (1974). Dougherty's main point is that case grammar so-called offers at best a possibly interesting but theoretically unimportant classification of verbs. Verbs, in this system, are classified on the basis of intuitive judgments about their complements. This–Dougherty goes on–might seem harmless enough, possibly comparable to classifying words by the number of letters, as in the *New York Times Crossword Puzzle Dictionary*; but it is not harmless, because Fillmore has deceitfully superimposed on this mere taxonomy the trappings of a generative grammar.

If the mere taxonomy charge is valid, then I should probably repeat that the deep case proposal was not intended as a complete model of grammar, but only as a set of arguments in favor of the recognition of a level of case structure organization of sentences. If the mere notational variant evidence is valid, this should not be taken as a reason for dismissing case theory, but rather as a challenge for considering such questions as which of the many possible notational variants of the final correct theory offers the greatest convenience for making generalizations about language typology, about lexical classification, about child language acquisition, and so on. In short, I see the notational and taxonomic possibilities of case theory as standing or falling together. A taxonomy is to be valued if it provides a convenient and revealing conceptual organization of the entities in its realm, in our case something in terms of which grammatical and semantic generalizations can be easily formulated; a notation is to be valued if it allows the formation of such a taxonomy in a simple and straightforward way.

The two sorts of objections I have just been discussing have been that case theory is *at best* a mere notational variant of some preferred theory, or is *at best* a mere taxonomy. The critics who have raised these issues have other objections, too, several of them relating to points discussed elsewhere in this paper.

But now let me turn to objections that require action. Stephen Anderson, in a paper called "On the Role of Deep Structure in Semantic Interpretation" (1971), challenged my claim that it would be possible to do without a level of deep structure in the sense of the Standard Theory. The position I had taken was that grammatical theory needs a level of case structure, which I thought of as being, or as being close to, a level of semantic representation; this position recognizes, by means of the grammar's transformational rules. the levels of surface structure. But, for reasons like those given by Halliday (1967, p. 39), there was thought to be no need for a separate level of deep structure.

I would say, for example, that in the case of the English verb *break* used transitively, its two arguments could have the case functions agent and patient, the agent being the entity responsible for the breaking, the patient being the entity which broke. In the surface structure, one of these arguments has to become the subject of the sentence. One possibility is for the agent to become the subject, as in 8a; another, which has side effects on the form of the verb, is for the patient to become the subject, as in 8b:

(8) a. John broke the vase.
b. The vase was broken by John.

I presented these as optional choices provided by a Subject Selection rule. The important point to notice is that, in the history of the formation of a passive sentence, no level of representation was posited according to which the sentence had a subject that was different from the surface subject. It had a case structure and a surface structure, but no intermediate structure within which the notions of subject and object could be defined.

Anderson, defending the Standard Theory, argued on grounds of the simplicity of the semantic component for the existence of deep-structure subjects and objects. His argument took the form of a claim that there are certain semantic generalizations which can be formulated simply only at the standard deep-structure level. There are two situations covered by Anderson's arguments. The first is that in which certain transformations, like Passive and Dative Movement, are viewable as changing an original assignment of deep-structure subject or object; the second is that in which a given predicate allows more than one way for its arguments to be parcelled out as terms of grammatical relations. In either case, Anderson's argument is that frequently there is a difference between a *holistic* and a *partitive* interpretation given to a noun phrase, depending on whether or not it has one of the primary grammatical relations (subject or object) at the deep-structure level. To illustrate Anderson's point from familiar examples, we can say that the difference between 9a and 9b is that the sentence with *the garden* in subject position conveys the idea of the whole garden containing bees, whereas in the other no such assumption is necessary; and that the difference between 9c and 9d is that the sentence with *the truck* in direct-object position conveys the idea of the truck being entirely filled with hay, an assumption not necessary with the other sentence:

(9) a. Bees were swarming in the garden.
b. The garden was swarming with bees.
c. I loaded hay onto the truck.
d. I loaded the truck with hay.

New examples suggested by Mellema (1974) are the sentences in 10:

(10) a. He read from his speech.
b. He read his speech.

Having in mind the reading-aloud sense of *read*, the sentence with *his speech* in direct-object position gives the impression that the entire speech was read, while the other sentence definitely does not.

Without feeling required to concede that these arguments support the version of deep structure that Anderson had in mind, I believe I must concede that a level of representation including the grammatical relations subject and object—or something else which recognizes the kind of close participation in the nuclear part of a clause that we find with subjects and objects—is probably necessary for grammatical theory. This does not mean, however, that a level of representation which recognizes case functions is spurious. We still have, after all, the problem of determining, from an understanding of what is being said, which of the arguments of a multiplace predicate are to appear as its subject, and which, if any, as its object.

The next truly worrisome criticism of case theory is the observation that nobody working within the various versions of grammars with 'cases' has come up with a principled way of defining the cases, or principled procedures for determining how many cases there are, or for determining when you are faced with two cases that happen to have something in common as opposed to one case that has two variants. In his careful paper "Some Thoughts on Agentivity" (1973), D. A. Cruse examined a number of linguists' statements about agentivity, found not only that different linguists disagree on the definition of agentivity, but also that the notions that enter into these definitions do not appear to be comparable enough to lend themselves to a uniform definition of a coherent concept.

I have thought of the problem of determining the cases as being somewhat analogous to that of determining the phonological units of language. Since *pie* and *buy* are different words in English, we have good reason to believe that we have a /p/ phoneme that is distinct from a /b/ phoneme. In *spy*, however, we have a problem. Should we analyze its stopped segment as /p/ or as /b/, or as something distinct from either of these? If it is not a /p/ or a /b/, should it be seen as something which in some sense includes both the /p/ and the /b/ phonemes, something which is exactly neutral between the /p/ and the /b/, or something conceptually distinct from those two phonemes altogether–something taking part in a different phonological system? The data do not dictate the correct answers to these questions. American phonemicists tend to regard the segment as /p/, but Danish phonologists, facing a similar problem in their language, choose /b/. Firthian linguists would tend to regard it as a systematically different item entirely; phonologists who allow certain kinds of abstractions might regard it as a category that includes both /p/ and /b/, and, if they make special assumptions about the feature composition of phonemes, they would recognize it as a segment containing features shared by both /p/ and /b/ but lacking any of the features which distinguish those two segments from each other.

An analogous array of positions could be taken on the problem of the identity of cases. Huddleston (1970) examines my arguments that in 11a *John* functions as agent and that in 11b *this key* functions as instrument. Here we are dealing with what might be called 'indirect cause' and 'direct cause', respectively. A reason for feeling sure that the two roles are distinct is that the same two nouns, preserving their case roles, can also occur together, with *open*, in a single sentence, as in 11c. The problem, Huddleston suggests, is what to decide about the causal element in 11d. Deciding on the case role of the subject in 11d is like deciding on the "emic" status of the stopped consonant in *spy*.

(11) a. John opened the door.
b. This key opened the door.
c. John opened the door with this key.
d. The wind opened the door.

One possibility is that the wind, like John in the earlier sentence, is using its own energy rather than energy provided by something or someone else, and that therefore it is the agent. A second way of looking at it is that, since the wind is the direct cause of the door's opening, it should be seen as the instrument. A third view is that the wind in 11d has a role that is distinct from both agent and instrument–call it force–since we have neither manipulator nor manipulated, but only a self-sufficient force. Another view, of course, is that there is a case feature cause that is *one* of the features of both agent and the instrument case; in the subject noun phrase for our sentence about the wind opening the door, the cause feature is present, but the features that distinguish an agent from an instrument are absent.

Since there are all of these possibilities to draw from–each sanctioned by one or another respectable linguistic tradition–there is little wonder that scholars who have tried to make the case theory work have not been able to agree. Furthermore, since linguists differ in the kinds of sentences they concentrate on and in the various positions that can be taken on the degree of closeness between case structure and semantic structure, and since not every linguist will think of the full range of possible alternatives, scholars working with caselike notions have all come up with different lists. The shortest list is John Anderson's (1971): 'nominative', 'ergative,' and 'locative'. The longest, I think, is that of William Martin (1972), who has separate case labels for the nouns found in the phrases *at the station, on the table, in the box, from the station, off the table, out of the box, to the station, onto the table*, and *into the box*. The principle seems to be that there are at least as many deep-structure cases as there are surface-structure cases or types of prepositional phrases.

Halliday (1967) recognizes a small number of different kinds of clauses, and assigns separate kinds of caselike structures to each of them. The method of L. Stephen Coles (1972), by contrast, is to divide verbs into a number of types (he finds sixteen), specifying a framework of caselike concepts for each type. With verbs of construction, for example, he distinguishes the resulting constructed object, such as the house that got built; the material out of which the thing got built; the people who performed the constructing act; the implements that were used; the time of the event; the location of the event; the time span from the beginning of the task to the end of the task; and so on.

In Coles's case, if I understand his purposes, there is no concern with whether the resulting-object case with a verb of construction is or is not the same as some particular case in the conceptual framework posited for some other verb class.

5.

I believe now that there might be a solution. It involves what I said earlier about meanings being relativized to scenes. One way for me to lead into this question is to point out that, as I have conceived them, the repertory of cases is *not* identical to the full set of notions that would be needed to make an analysis of any state or event. One of the cases in the system I proposed was the agent, identifying the role of an active participant in some event; yet *events* are not restricted in the number of active participants they can have. For example, in what I shall be calling the *commercial event*, two different individuals are agentively involved, and the actions of *each* of the two are part of our understanding of *any* of the lexical items that can be used for describing this kind of event or any of its aspects. The point I want to make here is that a case frame need not comprise a description of all of the relevant aspects of a situation, but only a particular piece or section of a situation.

One of the individuals in the commercial event–the buyer–hands over some money and takes the goods; the other–the seller–surrenders the goods and takes the money. A complete description of the commercial event would identify the buyer, the seller, the money, and the goods.[1] A prototypic commercial event involves all of these things, but any single clause that we construct in talking about such an event requires us to choose one particular perspective on the event. In the usage that I suggest, any verb identifying any particular aspect of the commercial event will constrain us to bring one or more of the entities in the event *into perspective*, the manifestation of this choice for English being the selection of grammatical functions corresponding to the notions of underlying subject and direct object. For example, if I wish to take the perspective of the seller and the goods, I will use the verb *sell*. Should I wish to take the perspective of the buyer and the money, I will use the verb *spend*. If I wish to bring into perspective either the buyer and the

1. A really careful analysis, in fact, would require that reference to the money be made from two points of view, the *cash* and the *value*. This would make it possible to distinguish:

(i) I bought it with a two-dollar bill.
(ii) I bought it for two dollars.

money or the buyer and the seller, I will use the verb *pay.* Should I wish to take the perspective of the goods and the money, I will use the verb *cost.* And so on.

In each of these cases, the speaker is required to construct a sentence in which one of the two or three entities that have been put into perspective becomes the subject: and, maybe, one becomes the direct object. The new question for the theory of cases is this: What do we need to know about the various participant roles in a situation in order to know which of these roles or which combinations of them can be put into perspective, and then, for those which have been put into perspective, which is to become the subject and which is to become the direct object?

The connection with the notion of 'scenes' can be stated this way. The study of semantics is the study of the cognitive scenes that are created or activated by utterances. Whenever a speaker uses *any* of the verbs related to the commercial event, for example, the entire scene of the commercial event is brought into play–is "activated"–but the particular word chosen imposes on this scene a particular perspective. Thus, anyone who hears and understands either of the sentences in 12 has in mind a scene involving all of the necessary aspects of the commercial event, but in which only certain parts of the event have been identified and included in perspective. The buyer and the goods are mentioned in 12a, the buyer and the money in 12b. In each case, information about the other elements of the scene could have been included–via nonnuclear elements of the sentence, as in 12c and 12d:

(12) a. I bought a dozen roses.
b. I paid Harry five dollars.
c. I bought a dozen roses from Harry for five dollars.
d. I paid Harry five dollars for a dozen roses.

One typical way of dealing with conceptually obligatory but superficially optional elements in a sentence is to claim that these elements are present in the deep structure but deleted or given zero representation on the surface structure. [Examples 12a and 12b represent different conditions for such deletion operations.] Within the view that meanings are relativized to scenes, however, it may not be necessary to believe that everything that is included in our understanding of a sentence is necessarily a part of the underlying grammatical structure of that sentence; it seems preferable to say that a word like *buy* or *pay* activates the scene of the commercial event; that everybody who understands the word knows what are the various components and aspects of such an event; and that a speaker's linguistic knowledge of the verb includes that knowledge of the grammatical ways in which the various parts of the event can be realized in the form of the utterance.

Any particular verb or other predicating word assumes, in each use, a given perspective. The grammatical functions of the nominals that represent the entities that are put into perspective are determined in part by something like a *deep case hierarchy.* Other parts of the associated scene can be introduced with prepositional phrases, with adverbials of various kinds, and with subordinate clauses. The "circumstantial" constituents of a sentence need not be aspects of scenes that are specifically required by a particular type of situation. Since any event takes place in time, any event sentence can contain a time adverbial; since many kinds of events take place in specific locations, sentences representing such events can contain locative adverbials; and so on.

Now when I say that meanings are relativized to scenes, what I mean is that we choose and understand expressions by having or activating in our minds scenes or images or memories of experiences within which the word or expression has a naming or describing or classifying function. For example, if we have occasion to say that a person has spent a certain amount of time *on land,* we know that this expression is chosen from a contrast set for which the opposing term is *at sea,* and that the terms from this set require a background scene involving in some way a sea voyage. Similarly, we know that if we describe somebody as being *spry* we have in mind a background scene of an age set for which the degree of activity and vigor we have in mind is relatively rare. The point is that, whenever we pick up a word or phrase, we automatically drag along with it the larger context or framework in terms of which the word or phrase we have chosen has an interpretation. It is as if descriptions of the meanings of elements must identify simultaneously 'figure' and 'ground'.

To say it again, whenever we understand a linguistic expression of whatever sort, we have simultaneously a background scene and a perspective on that scene. Thus, in our examples about buying and selling, the choice of any particular expression from the repertory of expressions that activate the commercial event scene brings to mind the whole scene–the whole commercial event situation–but presents in the foreground–in perspective–only a particular aspect or section of that scene.

6.

Languages, and lexical items, differ in interesting ways in the options they present in taking particular perspectives on complex scenes. Consider, for illustrating this point, a scene of a person taking something and causing that thing to come into abrupt contact with something else.

The English verb *hit* allows either of two perspectives on such a scene. One is that of the actor and the manipulated object; the other is that of the actor and the affected object. Illustrating the first of these we have a sentence like 13a; illustrating the second we have 13b:

(13) a. I hit the stick against the fence.
b. I hit the fence with the stick.
c. I hit the stick.
d. I hit the fence.

The event referred to in 13a and 13b may be the same, but the perspectives are different.

I will refer to the elements that are brought into perspective—the elements that appear as subjects and direct objects—as *nuclear elements* in the sentence. The first thing to notice about nuclear elements, so defined, is that they are not identical with the obligatory elements of the sentence. In 13a, for example, the prepositional phrase is obligatory, yet it is not a part of the nucleus. Put differently, 13d can be taken as an in-context abbreviation of 13b, but 13c cannot be interpreted as an abbreviation of 13a.

A question that must be asked is under what conditions something can be brought into perspective. In the case of a hitting act in which the two objects that come into contact are both inanimate and neither is described as being interestingly affected by the event, it is difficult to imagine any particular salience difference between the two, and hence it is not easy to see the difference between the two perspectival choices. However, just in case one of the two entities brought together in an act of hitting is in some sense inherently more worthy of inclusion in perspective, inherently more salient, the force of the perspective choice begins to become noticeable. One possibility for enhancing the saliency of the affected entity is to make it be, not a fence, but a person. We will find, I believe, that it is in some sense more natural to say 14a than 14b:

(14) a. I hit Harry with the stick.
b. I hit the stick against Harry.

This must be because it is more natural to include human beings within perspective than to leave them out in favor of inanimate objects. It is not that there is no use for sentence 14b, but rather that in that sentence the decision to leave Harry out of perspective has the effect of treating him as a physical object rather than as a sentient being affected by the hitting act.

For scenes involving mechanical actions of the kind just considered, it is fairly easy to set up a conceptual framework which has parts that match case-like notions in a fairly straightforward way. Thus, the thing which gets manipulated is the patient, the thing on which the manipulated thing acts is the goal, and the manipulator is the agent. What we can say about the verb *hit* and a number of other verbs within the same semantic field is that certain perspectives are available: agent and patient, as in 14a, with agent as subject and patient as direct object; agent and goal, as in 14b, with agent as subject and goal as direct object; and, with the manipulator left out of perspective, patient and goal, with patient as subject and goal as direct object, as in 15:

(15) The stick hit the fence.

In sentence 15, the decision to leave the manipulator out of perspective has the effect of taking a smaller perspective on a larger event; one situation justifying this choice would be that in which the stick was thrown into space, so that its coming into contact with the fence could be seen as a separately individuated event.

The verb *beat*, by contrast, is associated only with a scene in which the agent is holding onto the manipulated object throughout the action. Thus we can say 16a–d, given the special interpretation required for 16c; but we cannot use sentence 16e for the same scene:

(16) a. I beat the stick against the wall.
b. I beat the wall with the stick.
c. I beat the stick against Harry.
d. I beat Harry with the stick.
e. *The stick beat Harry.

Since *beat* necessarily involves an event in the history of the agent, there is no way of leaving the agent out of the perspective.

The verb *knock* offers a few special problems. Conceptually an act of knocking requires more than one participant, but only the agent need be put into perspective. Thus when we say 17 we have the agent's knocking action primarily in mind:

(17) He knocked on the door with his fist.

Thus we have seen how it is that the notion of perspective can be called on to explain (more honestly, to express) the subtle semantic differences, both in the grammatical organization of clauses having the same verb, and among sentences containing different verbs (*hit*, *beat*, and *knock*) within a single semantic domain.

One of the saliency conditions favoring inclusion in perspective, I have suggested, is humanness. Another is change of state or change of location. When an agent moves a patient against a goal, and as a result the goal participant moves or changes, the element in the goal case has acquired the saliency sufficient for it to be included in the perspective. We saw in 17 that with *knock on* the agent needs to be included in the perspective; but if in an act of knocking against a door, the door falls flat, we express this with the two-word verb *knock down* and include the door as direct object, as in 18:

(18) He knocked the door down.

Consider now the verb *push*, and suppose that I say 19a. Here we have a two-participant scene with a one-place perspective, meaning, if I am right, that it is the agent's action itself that is central. But now suppose that, as a result of my pushing against the table, the table moves. I can now say 19b, with the table in perspective, i.e. with *the table* in direct-object position.

(19) a. I pushed against the table.
b. I pushed the table.

The fact that some change occurred with the table is what made it natural to include it within the perspective.[2]

2. One possible interpretation of these phenomena, brought to my attention by students of János Zsilka when I recently spoke in Budapest about direct object functions, is that instead of speaking here of "choosing" to put some nominal in direct object function by virtue of noticing an enhanced salience, we should speak of a lexicalization process of transitivization or some other sort of valence change. Hungarian has many instances of sentence relationships of the sort I have been considering, in which the lexical difference is in the presence or absence of a perfectivizing prefix *meg-* (Zsilka 1967). I think that the observations about perspective that I have been making in this paper are correctly seen as involving the functions of nuclear grammatical relations; but I agree that in a linguistic description these notions belong more appropriately to an account of aspect and lexicalization patterns than to a description of the reasons speakers have for assigning a grammatical organization to their sentence. I have taken this more "processual" view more as an easy way of talking about the phenomena than as a claim about their place in the operation of a grammar.

Suppose, now, that I manipulate one thing, bring it into contact with another thing, and, as a result, that second thing moves or undergoes a change. The thing which changes, as we have seen, gets included in the perspective, leaving the manipulated object as outside the nuclear system of grammatical relations. It has to be marked, if it is mentioned, with the preposition *with.* The notion that I referred to in my earlier writings as the instrument can thus be seen as a derived notion, involving the relationship between an entity in one event and the event which is caused by that first event. (This interpretation of one part of the notion instrument has been made particularly forcefully in Talmy 1972.) Thus, if I hit a hammer against a vase and the vase breaks, I express that as 20a; if, however, I strike a hammer against a vase and the *hammer* breaks, I express that as 20b:

(20) a. I broke the vase with the hammer.
b. I broke the hammer on the vase.

In each of the two-event scenes, the first part involves one thing moving toward another; the second part differs from one scene to another. The scene in which the hammer breaks requires *the hammer* to be realized as the direct object, *the vase* appearing in a construction with a goal preposition; the scene in which the vase breaks requires the vase to be the direct object *the hammer* appearing in a construction with the preposition *with.*

According to this interpretation, the relationship between a change-of-state verb and the entity which undergoes this change of state is reflected, not in the underlying case structure (at least not in the underlying case structure of the first part of the two-event scene), but in the grammatical relation *direct object.* Having this in mind, we can now shed some light on the semantic difference between the sentences of 21:

(21) a. I cut my foot on a rock.
b. I cut my foot with a rock.

In the sentence with *with* the foot has the goal relation to the action, and the rock is treated as the thing which acted against the foot; in the sentence with *on* the foot has the patient relation to the action, and the rock is seen as the thing against which the foot moved. The thing which underwent the change of state–in each case, the foot–is expressed uniformly as the direct object, independently of its case role in the underlying action scene.

My new way of treating these observations–using both caselike notions and grammatical relations–requires, for verbs of the kind we have been examining, no separate independent instrument case, but instead a process of setting aside by means of the preposition *with* (in some cases, *of*) any patient noun which, by virtue of there being another noun phrase with higher salience, does not become part of the sentence nucleus. A result of this change is that there is no need for embarrassment about the fact that the same preposition has both instrumental and noninstrumental uses. With the sentences of 21, and with sentences like 22:

(22) I filled the glass with water.

the generalization for the selection of *with* is uniform: The entity marked by *with* is the entity which moved with respect to something else but was left out of perspective by virtue of the higher salience of its competitor, the goal entity.

This last point enables us to add to our list of saliency criteria. I have been proposing that the perspective taken in a clause is determined by some sort of hierarchy of importance, which might be called the *saliency hierarchy.* Humanness and change were the two conditions on this hierarchy that we have considered so far. I believe that definiteness and totality can be added to the list. With this decision we may be able to resolve some of the old problems with smearing mud and loading hay. In sentences 23a and 23b we have the idea that we are dealing with an event that more or less completely affects the truck or the wall–in particular, that the truck was filled or that the wall was covered as a result of the action. In the sentences that are typically paired with them, 23c and 23d, no such assumption is necessary:

(23) a. I loaded the truck with hay.
b. I smeared the wall with mud.
c. I loaded hay onto the truck.
d. I smeared mud on the wall.

The condition seems to be that if, as a result of bringing something into contact with something else, that something else is affected in some complete way, that new status of the goal entity is sufficient for its inclusion in the clause's perspective.

Some verbs allow a choice of perspective–for example, most of those we have looked at so far. Others, as we are reminded in Mellema (1974), have fixed perspectives. *Cover,* for example, requires that the goal be taken as direct object; and *put*, by contrast, requires that the patient be taken as direct object. These facts are illustrated in the sentences of 24:

(24) a. I covered the table with a quilt.
b. *I covered a quilt over the table.
c. I put a quilt on the table.
d. *I put the table with a quilt.

Lexically specific requirements of this sort must be included in the technical description of the words; just as with *beat*, as noted above, the continued relation between the one who does the beating with the implement he uses for the beating must be specified.

7.

What I have tried to do in the last part of this paper is to suggest that some of the problems of case analysis discussed in the earlier parts–in particular the vexing questions of the number and variety of cases–may be drawn a little closer to resolution if we separate from each other two things: one, the role analysis of the participants in a situation, of the kind for which proposals about deep cases have been made; and, two, the conditions under which a speaker can choose to draw certain situation participants into perspective. Since this perspectivizing corresponds, in English, to determining the structuring of a clause in terms of the nuclear grammatical relations, a consequence of this proposal is that it becomes necessary to recognize a level of grammatical structure which makes use of underlying grammatical relations, something I once thought it important to reject. I believe, however, that a level of case or role analysis is also needed, as a part of a general analysis of the scenes that get communicated with speech; and I believe that what unites these two sorts of structure is the notion of perspective that I have been trying to develop here. We recognize scenes or situations and the functions of various participants in these scenes and situations. We foreground or bring into perspective some possibly quite small portion of such a scene. Of the elements which are foregrounded, one of them gets assigned the subject role–in underlying or logical structure–and one of them–if we are foregrounding two things–gets assigned the direct object role in the clause. Something like a *saliency hierarchy* determines what gets foregrounded, and something like a *case hierarchy* determines how the foregrounded nominals are assigned grammatical functions.

6

Topics in Lexical Semantics

This long and confusingly named article was an attempt, in four lectures given at an LSA Linguistic Institute in the summer of 1976, to bring ideas of lexical semantics together with ideas about the larger task of knowing how linguistic texts are created and understood.

Since the case frames *in the earlier articles could be thought of as schematic underpinnings of situation types, and since the lexical items that bore these frames could have meanings that filled in the details of such schemas while at the same time evoking and depending on still larger kinds of knowledge structures, I found myself stepping beyond the boundaries of purely linguistic judgments about words and into questions of understanding, inference, and representations of event types that go well beyond anything that can be captured as constellations of semantic roles. The boundary between what we know about the language and what we know about the world seemed to disappear, but it could all hang on knowledge we have about the lexicon and grammar of our language.*

This led to two different levels of meaning description, one of which had to do with some understanding of "what was going on" – the states of affairs that people need to communicate with each other about; but the other was about the way in which the grammar of the language constrained the shape of the linguistic structures we could use for communicating. A situation could involve a great many participant roles, but the grammar of an English clause, and the ways in which individual words were coded to participate in the structures provided by that grammar, required speakers to make choices. The saliency hierarchy adumbrated in the preceding article is replaced with a list of preferences having to do with assumptions about the global nature of an entity's participation in an event, and not with the named "cases". These preferences had the effect of ranking the elements that could be included in the nucleus of a clause, so that for example the highest ranking element in the nucleus is the one that is chosen for the logical subject role. In some cases the "choosing" is done by the lexicon; in other cases it is

Originally published in 1977

done by the speaker, keyed to speaker judgments about the role or importance of particular arguments.

The article closes with demonstrations of how this liberated (undisciplined?) way of talking about meaning could be used to reconstruct a large number of traditional notions in the theory of linguistic semantics.

LECTURE ONE

1. My task in these lectures is to offer an informal and intuitive approach to the description of word meaning and text meaning. My main goal is that of presenting a uniform conceptual framework for discussing the meanings of words, the construction of sentence readings, the interpretation of texts, and the processes of expression and comprehension.

I believe that the linguist, in his consideration of a number of issues in semantic theory, can profit from the exercise of examining these issues within a larger view of language production and language comprehension. Previous traditions of semantic analysis have tended to be limited in their subject matter because of their commitment to lesser goals. Some linguists–Coseriu, for example–have gone to great pains to ensure that they are limiting themselves to what is exclusively and purely linguistic, free of contamination from knowledge about cultures, belief systems, or facts about the world. The anthropological or so-called cognitive semantics tradition has concentrated on the discovery and display of systems of discriminations in taxonomies, and their effort at finding the simplest representations of such systems has presumably been motivated by the desire to capture the culturally unique basis of a given taxonomy in a given linguistic community. Those structuralist traditions in which the emphasis was that of finding a core meaning for any given linguistic form seem to have had as their motivation that of accounting for the maintenance of a single linguistic form through time. In the generativist tradition, the concern has been largely a notational one, motivated by the desire to have, in the end, exactly the right number of features (or abstract predicates, depending on which branch of the tradition you are considering) that would simultaneously serve as counters in a system of semantic rules for the language–rules about truth and synonymy, for example–and as the basis for theorizing about the language-universal foundations of semantics. By contrast, the task I am taking on is that of locating a concern with meaning within a larger theory of language processing; the aspect that I will concentrate on here is that of text comprehension.

I would like to begin my discussion with an examination of the various steps in the comprehension process. One pastime indulged in by people working in text interpretation is that of choosing or inventing some text that offers particularly serious problems for the theory and challenging each other to show what can be done with it. The result of such demonstrations is usually–and justly–a sense of gloom about the prospects of an ultimate theory of texts. Since my interest, for now, is in discovering the steps in the interpretation process, without worrying too much about how in the end everything can be accounted for, I offer my example of a particularly troublesome text in good conscience. I do not pretend to be ready to talk about the final form of the correct theory.

My text is easy to understand; its complexity is in the description of the understanding process. It is taken from a recent issue of *Signature*, a magazine sent every month to holders of the Diners Club credit card. The article, clearly intended to be humorous, dealt with the public-speaking career of a professional baseball player named Pete Rose. It was mainly a collection of funny things that had happened in Rose's off-season public life. I will concentrate on just one sentence from the article, but I will tell you something about the preceding context.

Rose was giving a speech at an athletic club in Cincinnati. Since Cincinnati is Rose's home, his four-year-old son was in attendance. While Rose was speaking, unknown to him, the child moved up to the speaker's table and stood next to him. Rose told a joke, there was laughter, and after the laughter there was a moment of silence. Here, now, is the sentence I wish to bring to your attention:

> At precisely that moment a small but amplified voice notified the entire room, in the most lucid of idioms, of a common childhood emergency.

All mature native understanders of current English will know, of course, that the small voice was the child's; that there must have been a microphone at the table, which was responsible for amplifying the child's voice; that it was because of this amplification that the entire room got "notified"; that it was not, in fact, the "entire room" but rather the people in the room who heard the child s utterance; and that the author is not really using the word "notified" in strict literalness, given what we believe about the possible intentions of four-year-old children. We know furthermore, from personal experience, what this childhood emergency must have been; and from the phrase "in the most lucid of idioms" we even have a fairly good idea of the actual words the child might have used.

That is not all that we know about this piece of text. On another level, we have an idea why this episode was included in the article. That is, we know why people might consider it funny. We know what kinds of experiences are embarrassing; we know that when they are somebody else's experiences rather than our own, we find them amusing; and we know that the embarrassment is not serious, and that therefore our pleasure is forgivable if the embarrassing act is performed by a small child.

On yet another level, we know why the author described the episode as indirectly as he did. He did not come right out and tell us what the child said; he made us figure it out for ourselves. The effect here is accounted for by what Berkeley graduate student Robert Gaskins refers to as the Little Jack Horner theory of wit. Young Horner, on discovering the plum, was delighted mainly with himself, and only secondarily, if at all, with the baker. ("He put in his thumb, and pulled out a plum. And said 'What a good boy am I!'") If the plum had been sitting right on top of the pie, discoverable without any cleverness or effort on the diner's part, the experience could not have been half so rewarding.

2. As this exercise has taught us, whenever we are interpreting what somebody has said or written, there are four questions we have to answer for ourselves:

(I) What did he say?
(II) What was he talking about?
(III) Why did he bother to say it?
(IV) Why did he say it in the way that he said it?

Only the first of these questions–"What did he say?"–is among the traditional concerns of linguistics. Linguists professionally pay attention to what people have said, and they can more or less feel that their work is done when they have devised a system of categories and contrasts, a notation, and, possibly, a generative theory, which will enable them to make systematic statements about each part of what has been said.

The third question–"Why did he say it?"–brings us into the province of speech-act theory and the logic of conversation. (If the speaker said something, there must have been a reason for him to say it. He may have wanted us to know the information he was giving us, for example. If so, he must have assumed that we did not already know it. Since he said it at the time that he said it, it must have been relevant to something that was just then in the air.

Thus goes the kind of reasoning that one engages in in the various versions of speech-act theory.) The fourth question–"Why did he say it in the way that he said it?"–brings us to one of the concerns of rhetoric.

I wish to focus on the second question–"What was he talking about?" By that I mean, not simply what was the topic of the discourse, or of the portion of the discourse in question, but rather, what can we say about the "scene" or "history" or "situation" or "world" or "image" or whatever the speaker intended the hearer to create at this point in the discourse.

In giving an informal analysis of the interpretation process for our sample sentence, we might say the following. The preceding part of the text mentioned an athletic club banquet and identified our hero as the guest of honor, the after-dinner speaker; and we know something of what this must have been like from our own experiences in the world. A sentence in the text brought our attention to the child. We know about children's voices, so our awareness of the presence of the child makes it possible to know right away whose "small voice" the author identified. Our knowledge of microphones and their use helps us make sense of the reference to amplification; and we know that loud noises can be heard at greater distances, and hence potentially by more people, than soft ones. Our expectations about the possible intentions of four-year-old children and about the kinds of relationships a small child can hold with other people enable us to guess reliably that the verb "notify" in the text is being used in a nonserious way. Our experience of childhood embarrassments and their urgency, a part, surely, of our general belief systems about very young children, lets us readily know what the child's problem was; and our knowledge of alternative ways of talking about human body functions, and their different social valuations, makes us aware that some of these locutions are more straightforward, more "lucid," than others; and hence we can guess something about what might have been said on this occasion.

3. A question we need to consider at this point, of course, is whether there is any way of talking about such things within linguistics; that is, whether, within linguistics proper we have any business talking about settings, beliefs, expectations, memories of personal experiences, and the roles these matters play in the process of figuring out what somebody is talking about.

The argument that such questions are proper questions for linguistics can be based on the following facts: first, that some of the judgments and interpretation strategies that make up a part of this process are related in clear ways to choices of specific linguistic forms and categories; second, that there are sometimes differences from language to language in the ways in which linguistic material maps onto detailed images or experiences; and third, that it is possible to make use of many of these notions in formulating generalizations about the text interpretation strategies that people use. The arguments are, then, that the facts are lined with the choice of linguistic material, that they can vary from language to language, and that they sometimes permit summarizing generalizations–matters that linguists, if not "linguistics" in some narrowing definition, ought to be able to deal with.

We make use of many kinds of judgments and awarenesses whenever we interpret a language text. We need to be aware of the nature of the ongoing communication act. We need to make use of our knowledge of whether the communication is urgent or playful, whether or not it is accompanied by actions on the part of the participants, and so on. In the present case, for example, we make critical use of the knowledge that the text is a report of supposedly humorous events, is intended to entertain, and is meant to be read at leisure.

At any point in the discourse the interpreter needs to be aware of scenes or images or memories that are, so to speak, "currently activated," a point argued at some length by Wallace Chafe in his recent work on consciousness and language (Chafe 1974). We were told early in the *Signature* article that the child was at the banquet; just before our target sentence, however, we are told that the child was standing next to Rose, and that Rose was unaware of his presence. We have in our minds a kind of outline image of an after-dinner speech setting, with the relatively clear parts out of the outline being the father and son standing next to each other.

We sometimes have to appeal to memories of experiences, sometimes to ongoing experiences, sometimes to procedures for acquiring experiences. Memories of inconvenient childhood experiences and the attendant sense of helplessness and dependence have to be available to us for understanding what is going on in this story. Awareness of current experiences is necessary for the understanding of sentences like "Put this one over here and that one over there." We have to notice what the speaker is doing while he is talking in order to know what he is telling us to do. And procedures for acquiring experiences are necessary for understanding a sentence in James McCawley's cookbook, which instructs the reader to knead a certain pastry dough "until it has the texture of an earlobe."[1]

We often have to have knowledge of what a number of workers in artificial

1. Personal communication, George Lakoff (who owns an advance copy of the McCawley cookbook).

intelligence call "scenarios" or "scripts,"[2] conventional or familiar routines or behavior sequences in terms of which we analyze and individuate larger events by seeing smaller events as parts of them. We have a scenario, for example, for the kind of after-dinner speech that is most likely to be given before an athletic club. We know that most commonly it would consist of, or at least include, lots of jokes, that the jokes would be followed by pauses during which the audience would laugh, that the general mood, for the ideal after-dinner speech at any rate, is one of jovial good humor, and so on.

And we have to have available to us certain strategies for constructing, from the pieces of a text, from our knowledge of the world, and from our estimates of the author's or speaker's purposes, some single coherent view of what is going on, some possibly complex but unified scene or story or world that we would recognize has been matched to this particular text.

4. This series of lectures, somebody is surely eager to remind me, has been given the title "Topics in Lexical Semantics." My emphasis, in fact, will be on the use of some of the notions I have been talking about in the description of individual lexical items. Lexical items will have to have their meanings described, if I am right, in ways that will show something of their contribution to a process of the kind I have been talking about. I can begin to show you what I mean by that by examining some of the things we have to know about the English verb *write* in order to understand sequences containing it, knowledge on the basis of which we can construct text meanings and make judgments about text coherence.

The verb *write* in what we might call its "prototype" meaning depicts an activity in which somebody is guiding a pointed trace-leaving implement across a surface. I suggest–and I think that there will be a general agreement on this–that this prototype scene associated with the verb is more or less what is basically understood when other information is not provided. Skywriting and writing in the air with one's fingers are departures from this prototype and would be understood as the intended scene for a text only if there was explicit information to that effect in the text or if the interpreter had very special knowledge about the context.

The prototype scene associated with *write*, then, contains the individual that does the writing, the implement with which the individual writes, the surface on which the writing is done, and the product of a writing act–that is, some configuration of marks on the surface.

---

2. See, e.g., Abelson (1973), Charniak (1974), Minsky (1975) and Schank (1975).

If the only thing we know about a text is that it contains the verb *write*, then the only associated scene we can construct for it is a kind of simple outline scene with positions for the entities I mentioned but with none of the details filled in. We never, of course, have only one word to go on: the word is embedded in a text.

Suppose that I say to you, "Harry has been writing." You will know that the scene I have asked you to create is being presented as an event in the real world, and that I expect you to know who Harry is. And so the writing scene you create in your mind is one in which Harry is the one who is doing the writing, the one who is guiding the trace-leaving implement. A part of this depends on the fact that I used a progressive form. If I had said, "Harry wrote," you would feel that I had not quite finished saying what I had planned to say, because with the preterit we have the idea of a completed act rather than a time-extended activity, and you have the right to expect some indication of what it is that Harry wrote. If I had said, "Harry writes," you would have the right to assume that I thought I was answering a question about what Harry does for a living. As these observations indicate, the knowledge that users of a language have about individual words may create different scenes for different linguistic contexts–differences illustrated here by choices of tense and aspect. And, elements that are within specific grammatical relations with the word–such as the subject of the verb–contribute in specific ways in providing details to the scene that the interpreter constructs for himself.

It may well have been, of course, that the knowledge that we already share about Harry will give you a framework for fitting in this new information about Harry's having been writing. Suppose you know, for example, that Harry is in the middle of his dissertation and that he always does his writing at a typewriter. Then, if you know a few other things about him, it might be quite appropriate for you to give the responses that show up in the following dialogues: I say to you, "Harry has been writing," and you say, "Has he finished chapter seven yet?" Or I say to you, "Harry has been writing," and you say, "I thought his typewriter was broken." These second-speaker responses can be seen as showing how the outline scene presented with the sentence can get fitted into specific kinds of belief-worlds on the interpreter's part.

If we do not have ready-made scenes into which to integrate the outline scene provided by a text or piece of text, then there are many "positions" in the text-associated "scene" that are left blank, so to speak. I believe that one way of exploring the meaning of a word is to consider what questions are relevant at the point when we have received and processed the latest bit of text, and the extent to which these questioning possibilities are determined by the presence and the selection of the word in question. The questions of the sort I have in mind are those designed to fill in the blanks of the outline scene that is in the process of being created. If *write* has the interpretation I have given it, then it follows that we would need very special contexts before we could believe we were talking coherently in the following examples: Somebody says to me, "Harry has been writing," and I say, "What time is it?" or I say, "When do you think I should go home?" Questions that do make sense, however, are questions like, "What did he write?" or "What was he writing on?" or "What was he writing with?" These are questions that identify positions in the schematic scene activated by *write.*

The points that I have made so far, I am sure, do not quite ring true, and that is because something has been left out of the description of this verb that is absolutely essential to its interpretation. The prototype scene associated with writing does indeed involve somebody guiding a trace-leaving implement across a surface, but it carries more with it than just that. Whenever we use the verb *write,* it is also understood that the *product* of the act of writing is something *linguistic*–that is, is something that represents linguistic forms. Since that is a necessary part of the scene associated with *write,* it follows that we could also ask, coherently, when told that Harry has been writing, such questions as: "What language was he writing in?" or "What does what he wrote mean?"–in short, we can ask about anything that is a necessary part of any scene in which a sample of written communication is a part. This last aspect of *write* distinguishes the English verb from the Japanese verb *kaku,* which is the usual–and usually correct–translation of *write.* The Japanese verb does not dictate the same assumptions about the nature of the product. The result of an act of *kaku*-ing can be a word or sentence or letter, or, just as well, a painting or sketch or doodle.[3]

I suggest that the first two tests I proposed for the relevance of talk about text/scene correspondences are satisfied. The first was that the choice and character of particular scenes could be related to the choice of particular elements; the second was that languages could be shown to differ from each other in ways that can be made clear by appealing to scenes and properties of scenes. For the first of these there was the choice of the verb and a given tense and aspect, with a particular grammatical subject; for the second we have the difference between the English and the Japanese verb.

5. The slogan that goes with the approach to semantics that I am taking is this: Meanings are relativized to scenes. I am using the word *scene* in this dis-

3. It should be pointed out that the Chinese characters used in writing the word in these two senses are distinct.

cussion in a technical sense that includes its familiar visual meaning, but much more as well. I mean by it any coherent individuatable perception, memory, experience, action, or object. Some scenes are made up out of other scenes; others cannot be analyzed, but must merely be known, that is, can only be demonstrated or experienced, not explained. The point I am making is that for a great many words and phrases in our language, we can only understand them if we first know something else, and this something else may not be analyzable. If you know what birds look like, then I can identify a certain part of a prototypic bird and tell you that part of it is called its "beak." To understand what is meant by such verbs as *wink*, *crawl*, *sneeze*, *yawn*, etc., you have to know about bodies–especially human bodies–and what kinds of things their owners can do with them, and what kinds of internal things can happen to them. To understand what is meant by *heartburn* you have to have had certain kinds of body experiences, to understand what is meant by *déjà vu* you have to have had certain mental experiences.

A word appearing in a text that is being interpreted by someone who understands the word can be thought of as activating a scene and pointing to a certain part of that scene. *Beak* induces the scene of a bird, and identifies a particular part of the bird.

So far my "scenes" have been static, or nearly so. For many kinds of verbs, or for adjectives indicating temporary states, the word will focus on a single period or stage of a history, but this single period must be understood as a part of a larger scene. It is as if we are looking at a single frame in a filmstrip, but our view takes in, at least in outline, the rest, or part of the rest, of the filmstrip too.

With respect to this last point, consider the following situation: Two identical twins–let us call them Mark and Mike–are patients in a hospital. They are staying in adjacent rooms, and as we walk by their rooms, accompanied by their nurse, we notice that each one is seated on the edge of his bed, in exactly the same position. By coincidence they have both assumed the same postures and they are wearing absolutely identical gowns. Photographs of the two scenes could not be told apart.

The actual visual scenes, then, are indistinguishable. Yet, as the nurse walks by Mark's room, she says, "I see that Mark is able to sit up now"; and as she walks by Mike's room, she says, "I see that Mike is able to sit down now."

To say of Mark that he is sitting up places the scene we have just observed in a history, and a preceding period of that history has Mark lying down; to say of Mike that he is sitting down places what we have observed of Mike in its separate history, and in a preceding period of that history, Mike was standing up.

If we know everything about Mark and Mike that the nurse knows, then we will take her utterance as merely expressing a judgment that we might have made too. If, however, we are merely overhearing her remarks, then we will probably find ourselves exercising our text-processing strategies by passing to the question "Why might she have said that?" We guess that the boys are in the hospital because of some kind of health problem. We know that in a hospital it is more or less taken for granted that the patients' conditions will change, and it is hoped that their conditions will change for the better. These bits of knowledge induce us to create larger scenes in which we see the present scenes as parts. We might conclude that the present positions of these two boys represent improvements in their conditions; and that would get us to believe that Mark was sick with something that made it difficult for him to sit up, and Mike with something that made it difficult for him to sit down. We end up with fairly detailed, but still in outline only, opinions about the hospital stays of these two boys and of the significance of their present states. When later on we learn that one of them has been suffering from mononucleosis and the other from hemorrhoids, we have ready-made frameworks within which to integrate this new information, and we are not likely to be confused about which boy had been suffering which of the two disorders.

Here is another example. The same two brothers, many years later, spend a few hours in San Francisco, separately, but doing essentially the same things. Later that day, Mark writes home, "I spent three hours on land this afternoon," and Mike writes home, "I spent three hours on the ground this afternoon." The actual observations we might have made of these two boys while they were in San Francisco could not have given us a basis for choosing these particular descriptions, yet we know that one of them expressed the outing as an interruption of a sea voyage and the other as an interruption of a period of air travel. We know this because we know that for the sea voyage scene English provides a pair of expressions, "at sea" and "on land," for a particular two-way contrast. Since this is a contrast between mutually exclusive alternating states, since we know that Mark is talking about himself, and since the sentence about his having spent three hours on land is in the past tense, we can assume that he was once again at sea when he wrote the letter. Similar reasoning is appropriate for the two-way contrast "on the ground" and "in the air," expressions relativized to scenes of air travel.

Once again, the comprehension process involved knowing what larger scene the expressions were relative to, what parts of those larger scenes were being focused on, and what further "computing" could be done in creating the larger scene, by considering such things as the use of tenses.

6. I need to introduce the concepts that I think are necessary for discussing the communication and comprehension processes. I am still finding it difficult to give precise definitions of these notions–and I am appropriately embarrassed by the conclusions that *that* state of affairs invites–and so I will merely try to show you some samples of how I talk about meaning and comprehension.

The meanings of individual lexical items can best be understood in terms of their contributions to the process of interpreting a text. This involves, as I have suggested, much more than the processing of meanings directly provided by the text, but also memories, knowledge, and current perceptions of the interpreter, as well as the application of a set of procedures for determining the basis of the coherence of the text. A word or phrase or sentence or text identifies a scene, and it foregrounds, or highlights, some portion of it. It is to be understood that the identity of a scene can be established at any number of levels; an event, for example, can be seen as composed of a number of subevents and as being a part of some larger event or situation; I allow myself to use the word *scene* for each of these.

Sometimes we can be said to have a scene at the ready, made up of our general knowledge, memories, imaginings, and whatever is currently a part of our conscious awareness; and we can also speak of constructing a scene, and in the act, making sense of what we are reading or what somebody is telling us. The scenes to which the meanings of linguistic forms are said to be relativized are individuated in ways that are partly natural, partly convention, partly idiosyncratic. At any rate, we can speak of the individuatability of scenes in terms of their fit with certain prototypes or paradigm instances. (Much of the work of the Berkeley psychologist Eleanor Rosch can be thought of as being directed toward the discovery of the mechanism of scene formation in my sense; her early research concerned prototype notions of natural kinds, colors, shapes, and basic level artifacts; she is currently considering the question of the ways in which the character and the boundedness of events and situations are determined in the categorization process (Rosch 1973).)

A given real-world scene is perceived according to the degree to which it matches some paradigm or prototype outline scene. Prototype scenes can be thought of as scenes from simple worlds, worlds whose properties simply do not take in all of the facts of the real world. Prototype scenes account for the clearest cases, the best examples. Much of our use of language, however, requires us to use words that activate a prototype scene even though we are talking about something that departs from the prototype. A widow is a woman whose husband has died. The prototype "scenario" for widow does not cover the case of a woman who murdered her husband, or of a woman who had three husbands and now has only two left, or of a woman whose divorce became final on the day of her husband's death. If we do not know whether to use the word *widow* in these cases, it is not because we are unsure of the meaning of the word, but because the prototype scene that provides the setting for our knowledge of the word *widow* simply does not cover all these cases.

A final concept in this survey is what I will be calling "perspective." At any point in a text we are viewing a scene from a particular perspective; that is, while we have the whole scene in view, so to speak, we are focusing on just some portion of it.

The words I intend to use in this discussion, then, include *scene*, *prototype*, *activation*, and *perspective*. A few others will be introduced in some of the later lectures.

7. Earlier I identified a scene in which somebody was acting in such a way as to leave traces on a surface. The English verb *write*, I pointed out, is one that adds to that scene the information that the product is something linguistic. The word *sketch*, by contrast, adds the information that the product is something representational–a sketch, drawing, or diagram of something–and includes the information that the implement used is of a particular kind: it is not, for example, a brush. *Draw* seems to be more or less like *sketch*, but both are different from *paint* since to *paint* (in the use with which the direct object names the *product*) is to produce an artistic object, but not necessarily something that is representational. And the verb *paint*, as we know, brings special assumptions about the medium. *Printing*, as it is closely related to our prototype scene–as opposed to what one does with a printing press–requires the understanding that the product is made up of alphabetic elements of a particular form. There is a general verb *write*, but in two-way contrast between *printing* and *writing* the word *write* refers to the act of producing a product of the cursive or nonprinted form.

Only when the product is a representation of the writer's name, in a written rather than a printed form, and in an individualized form, can the verb *sign* be used. Further, only when the action is that of producing a sample of one's signature for the sake of somebody who wants it as a souvenir can the verb *autograph* be used.

The product of an act of writing can be named with a noun, and when this is done a new scene is introduced, the scene associated with the product. It can be the name of any of a number of linguistic-product units, such as word, phrase, sentence, paragraph; or book, article, poem, essay; or letter, preface, dedication, obituary, or confession, and a very large number of others; or it can be a *signature*, that is, an individualized rendering of the writer's name, or an *autograph*, that is, a signature that has been written as a memorabilium.

There are verbs that increase the detail of the dynamic aspect of the scene, that bring along some understanding, for example, of the manner of writing. These will include *scribble* and *scrawl*, which indicate a messy or careless manner of writing; *print*, which indicates a careful manner of writing plus the notion mentioned earlier of the nature of the product; and *jot down*, which indicates the brief recording of a short text.

Other verbs make specific the kind of implement that is being used. This is true of verbs taken directly from the names of the implements, such as *pen* and *pencil*, as well as the verb *type*, which indicates the use of a typewriter.

Some verbs activate scenes with the necessary presence of a source or model of what is produced. With *transcribe* the model is spoken; with *copy* the model is graphic, and not necessarily linguistic; with *trace* the model is understood as being directly beneath a translucent surface.

For all of these verbs the grammatical subject in the underlying form is the Agent, the person who performs the trace-leaving act. In many cases the direct object is some description of the product; thus: "He penned his answer," "He scribbled a poem," "He drew a picture," "He signed his name," and "He wrote his name." In the case of *write* it is possible to use the verb without an explicit product name if the product is understood to be a letter and as long as it is understood that the letter its intended to be sent to some person; in that case the indirect object phrase can stand as the only explicit complement to the verb. Thus, in "She wrote to me twice" or "I'll write you more about this later on," it is understood that the product of the writing is a letter. This choice is not possible for "he drew me" or "he scribbled to me" or any of the others we have considered.

In the case of *sign*, the direct object can be the name of the document that in some way becomes official as a result of the signing act. In this case the product is understood to be a signature, but a further aspect of the total scene is that the writing of a signature onto a particular document made that document valid. Thus I can say, "I signed the contract," "The king refused to sign the treaty," and even—since an autographed book is something special—"He signed my copy of his book." Notice that in each case the product is the writer's signature.

The verb *autograph* is one with the notion of signature built in. In the case of *sign* you can say "He signed his name" but not "He signed his signature"; with *write* you can say either "He wrote his name" or "He wrote his signature."

In the case of *autograph*, however, there is no way of expressing the nature of the product, but only the name of the object onto which the signature was written. Thus we can say "He autographed my football" or "the menu" or "my copy of his book."

Sometimes the purpose of writing one's signature is extremely specific, and in that case there may be a separate technical term just for the act of signing. An example is *endorse*, as in the phrase "endorse a check."

One highly specific writing-act verb is *address*, which can take as its direct object only nouns naming things that can be mailed, like letters, packages, and postcards. It means, of course, to write on each of these the destination of the thing. *Addressing* something is specific not only with respect to the content of the product, but also with respect to the purpose: thus I am not *addressing* an income tax form or a job application when I fill in my name and address.

8. These examples tell us something of what is required of a notation system for the lexicon. We need to know for each word what scene, or cluster of linked scenes, is to be activated by it; how, with a given meaning relative to a given scene, it is to be combined with other lexical elements, and what grammatical relations these will hold with each other. And this, as we have seen in the simple case of *write*, must have added to it some account of the effect of the choices of particular tenses and aspects. For example, there is a sense of *write* which designates an activity; it is this that is compatible with the progressive aspect. There is another sense of the verb in which it designates an act; it is this that is compatible with the preterit tense. There is another sense of the verb in which it designates a profession or avocation; and this is compatible with, among others, the simple present tense. This sort of information, I suggest, is to be given in connection with the description of the linguistic choices available for individual scenes. From that it will follow, through an understanding of the semantic force of the tenses and aspects, which of these are compatible with which senses of the verbs.

The theory of the lexicon will specify for each verb, with respect to each scene to which it is relevant, what we might call, following Tesnière and a number of European lexicologists, its "valence," the concept of valence being extended to include simultaneously reference to the surface form of sentences containing the verb and to the elements and aspects of the associated scenes (Tesnière 1959 and Helbig and Schenkel 1973). In the writing scene, for example, there are the writer, the implement, the surface, and the product. For all of these verbs, the writer appears as the subject. For some the product is realized as the direct object; for others the direct object is the name of the receiving surface. The implement is introduced into sentences related to the writing scene, if at all, by means of a *with*-phrase. For a number of special scenes, in which the writing act serves some larger purpose having its own, dominating scene, the structuring of the sentence can depend on facts about this larger scene: here I have in mind the indirect object structure with *writing* understood as *letter writing*.

Another part of the details of a "valence" description corresponds to what are sometimes called "selection restrictions." That is, some words are limited to highly specific scenes, and for that reason the other words in the structure have to be of a limited sort. I would introduce such information as a description of the scene rather than as a special kind of relationship holding between words or word classes in given grammatical constructions. I will have more to say about this in the next lecture.

9. In later lectures I will say something about the semantic role of subjects and objects associated with a verb; the relative saliency of aspects of scenes that determine which parts of a scene get realized as which kinds of grammatical constituent; the relation of lexically determined scenes to immediate context, and hence to questions of deixis and pragmatics; the performance of this kind of semantics in the description of traditional semantic notions; a survey of lexical domains; and some concluding predictions and confessions about the future of semantic research within this framework.

LECTURE TWO

1. Jeffery Paige and Herbert A. Simon, in their paper "Cognitive processes in solving algebra word problems" (1966) have pointed out three different strategies by which students interpret and 'solve' algebra problems formulated in words. I believe that from their results we can see something of the ways in which ordinary discourse is understood. This will be a fairly clear illustration, though with a slightly unnatural text, of some of the points I was making in Lecture One.

Here is an example of one of the problems discussed in their article. Pay close attention.

> A board was sawed into two pieces. One piece was two-thirds as long as the whole board and was exceeded in length by the second piece by 4 feet. How long was the board before it was cut?

The various strategies that Paige and Simon found students using for solving this problem drew differently on the scene-constructing principles I spoke about in the first lecture.

Students' responses to this problem were of four kinds, one being, of course, to give up. The three more interesting responses came from people who, in their opinion at least, knew what to do. Students in the first of these groups gave the answer as *minus twelve feet.* They formulated the set of relationships given in the text as a linear equation, with x as the unknown length of the original board, and then solved for x. One of the pieces, the text tells us, is two-thirds the length of the original board; its length is $2x/3$. The other piece, then, is $x/3$. Since the piece that is two-thirds the length of the original board is four feet shorter than the other one, the equation we need must be

$$2x/3 + 4 = x/3$$

Solving for x gives us -12.

The students who followed this first strategy did what they learned to do in mathematics lessons, and they were able to do so by suspending any inclination to interpret the text in an ordinary, everyday way. Mathematicians, of course, need to develop precisely that ability. A young student who took this first approach may become a fine mathematician some day, but maybe not a particularly good engineer.

The second strategy was to take the text as a description of some possible scene, and to see how things work out in the mentally constructed world containing this scene. This process cannot be carried out, since it requires that there be two board segments each of which is longer than the other. The response made by students who used this strategy was that the problem was unsolvable. The problem supposes things that cannot be. The user of the second strategy is a careful reader and a visualizer. He may become a good philosopher. In philosophy you have to learn to read carefully and to consider whether what you read is believable.

The third strategy is to assume that the teacher or experimenter is presenting the problem in good faith, free of tricks and deceit, and therefore assume, perhaps without even noticing the contradictions in the text, that the words "is exceeded by" were intended to be "exceeded." The answer given by the follower of this strategy is *twelve feet.* The strategy is to figure out what the speaker is driving at, to construct the scene you think he wants you to construct, and to proceed accordingly. A person who does so—whom we can call a cooperative visualizer–will make a good engineer. A good engineer can figure out what you need even if your description of what you need is flawed.

It is this last strategy–one that in this case gives a "wrong" answer–that is characteristic of the kind of comprehension process that is most typical for language understanding in normal settings. You hear what a person says, you try to figure out what he is talking about, you guess what his reasons are for saying what he is saying, and you let these beliefs influence your interpretation of what he said. (You may even let these beliefs influence your perception of what he said.) Actually, of course, the person who gives the correct ("unsolvable") answer uses this third step, too. In his case, instead of assuming that the interaction is completely in good faith, he assumes he is in a testing situation and senses one of the things he is being tested for is his ability to read carefully.

2. The scenes we construct for texts are partly justified by the lexical and grammatical material in the text and partly by the interpreter's own contributions, the latter being based on what he knows about the current context, what he knows about the world in general, and what he assumes the speaker's intentions might be.

The lexical information we need in a description of the workings of a language includes more than information about the nature of the associated scenes; it includes information about the grammatical form of the sentences in which the lexical item is free to appear. In particular, if the lexical item is a verb, we need to know which of the possibly many separate entities in the associated scene is to get realized as the subject of the verb, which is to appear as the direct object (if there is to be one), and in what grammatical guise the others will appear.

Some years ago I came up with a proposal (Fillmore 1968) on how this problem should be dealt with. It was in the form of a representation of sentence meanings that was deeper than deep structure, which had labels for the roles entities could have in a situation or event–roles like Agent, Patient, Instrument, Goal, Experiencer, Location, and so on. It defined a ranking or hierarchy among these notions, together with a set of principles that I referred to as the Subject Selection Rules. I referred to the role notions as deep structure cases; I claimed that any clause could be represented as a verb, indicating the nature of an event or situation, together with a collection of noun phrases marked with indicators of the case roles of each entity realized as the noun phrase; and I hoped that the subject-and-object structuring of a sentence, at the surface level, could be provided by rules of subject formation and rules of direct object formation. What I proposed assigned one kind of organization to sentences at a semantic or conceptual level, another kind at the surface structure level, with no intermediate level corresponding to Chomsky's 'deep structure' notion.

Of the many pages that have appeared in the journals and in my mail deploring Case Grammar, which was what I called this system, the one that hit hardest was a piece written by Stephen Anderson on the semantic role of deep structure (Anderson 1971). Anderson pointed out a number of semantic generalizations that could only be given a simple formulation at a level of deep structure, since they involved the semantic roles of deep structure subjects and objects. Anderson's point was that noun phrases that function as subjects and objects in deep structure have holistic interpretations that are missing in nouns that are sentential complements or adjuncts other than subjects or objects. I will have occasion to mention some of his examples later on.

In the meantime, an underlying structure for grammar has been proposed by Paul Postal and David Perlmutter and by David Johnson (Johnson 1974a, b and Perlmutter and Postal 1974). In it the notions subject, object, and indirect object are to be taken as primitive grammatical relations, assigned initially in a fixed way to underlying representations and modifiable by rules. They call their system Relational Grammar. In their terminology, elements that are subjects, objects, or indirect objects are called Terms (other noun-phrase constituents of clauses being Non-Terms), and terms are given ranks–namely, 1, 2, and 3 for Subject, Object, and Indirect Object, respectively. They have proposed a great many generalizations about grammar that they claim can be formulated most easily as operations on terms; these operations include the reranking of terms and the removal of termhood. Their system, too, recognizes a distinction between underlying subjects and objects and surface subjects and objects. Other scholars, especially Edward Keenan and Bernard Comrie (Keenan and Comrie 1972 and Keenan 1975), working independently, have proposed a number of typological principles as well as a great many language-universal processes and constraints on grammatical rules in which the notions subject, object, and indirect object play a central role.

I find much of this work compelling, and I am willing to reintroduce something like the notion of deep structure representation–at least a notion of the assignment of subjects and objects in a particular way to underlying predicates. Anderson's observations can, of course, be formulated in a relational grammar; that means that the form of grammar we ultimately decide on does not have to be one in which grammatical relations are defined solely in terms of tree structure configurations, which I take it as what Anderson himself, following Chomsky, would prefer.

3. Since, as I have been arguing, we can talk about the scenes underlying a text–in particular, the scenes underlying single-clause sentences in a text–we ought to be able to examine the principles according to which aspects of scenes map onto grammatical representations and determine the subject/object structure of clauses.

I will refer to the part of a sentence that contains a predicating word and the grammatical terms that are linked with it as the Nucleus, the remainder as the Periphery, of the sentence. We need to ask two questions concerning the relationship between scenes and the clauses that activate or are activated by these scenes. The first is, "Which of the entities in a scene get realized in the associated sentence as members of the nucleus?" Simplified, the question is: "What enters the nucleus?" The other question is, "When two or more elements from a scene get realized in the associated sentence as members of the nucleus, are there general principles that determine which of these is the subject, or first term, and which is the object, or second term?" (I disregard here the indirect object, or third term.) Put differently, our second question is, "What determines the ranking of terms in a nucleus?"

It is conceivable, of course, that the answers to these questions will be different for different languages. If that turns out to be the case, we must ask ourselves to what extent the terms of relational grammar can be thought of as semantic primitives, which is what Postal and Perlmutter have in mind. Can it be that language-universal properties–of the kind Postal, Perlmutter, Keenan, and Comrie have been suggesting–are to be found in categories that are themselves defined in language-specific ways? On its face, that possibility is not ridiculous, since apparently kinship semanticists find universally valid principles across kinship systems even though the most basic relationship within such a system, the parenthood relationship, may be conceived of differently in different cultures.

4. Answers to the first question involve mainly the problem of the semantic import of the direct object function. English requires every sentence to have a subject, and so the question of whether or not a sentence will have a subject is less interesting than the question of whether or not it will have an object. And this question, obviously, becomes relevant only in the case in which it is clear from the meaning of the verb that there are two entities in the associated scene, one of which must become the subject, the other of which may or may not become the direct object.

Jespersen (1927, p. 230) writes that "The notional relations between verbs and their objects are so manifold that they defy any attempt at analysis and classification."[4] The task of finding generalizations about the semantic import of the direct object is a difficult one, to be sure, but maybe not a hopeless one as long as we are willing to see the problem in a slightly different light. I suggest that we ask, not what kinds of relationships there are that link objects with their verbs but what properties of scenes determine whether or not something will be realized as a direct object. Some of the examples that most interested Jespersen can be thought of as resulting from fairly superficial amalgamational processes and can probably be ignored. I, at least, would not particularly trouble myself with sentences like "She laughed her thanks" or "He nodded his acquiescence" until I see whether they can be made to fit principles that seem to govern the clear cases.

5. In exploring the second question, we should consider four situations: (1) That of verbs that can realize either of two things as direct object, the leftover element becoming a peripheral element, expressible in a preposition phrase. An example of this first situation is *blame*: We can say either, "He blamed the accident on me" or "He blamed me for the accident." (2) That of verbs that can realize either of two things as direct object, but with constraints on whether the leftover element can be expressed at all. An example of this second kind is the verb *sign*. We can say, "He signed his name" and "He signed the contract"; but while we can say, "He signed his name on the contract," we cannot say, *"He signed the contract with his name." (3) That of verbs that can include a nonsubject noun either in the nucleus or in the periphery. An example is *shoot*: We can say either, "She shot him" or "She shot at him." (4) That of pairs of verbs that have related meanings but that differ from each other in the choice of what is in the nucleus and what is in the periphery. Verbs exemplifying the fourth situation are the pair *put* and *cover*: We can say either, "He put a towel over the alarm clock" or "He covered the alarm clock with a towel." Expressing the same situation, *towel* is the direct object in one sentence, *alarm clock* in the other.

The question I am asking is different from, but possibly related to, the question of the reasons for creating *surface* subjects and objects. Surface subjects can be created by passivization, raising into subject position, applications of tough-movement, and so on; and surface objects can be created by dative movement, raising into object position, reduction of range complements, and so on. The question I am asking is what assignment of terms is available to lexical verbs prior to any application of grammatical processes of

4. J. Otto Jespersen, *A Modern English Grammar*, vol. III (London: Allen & Unwin, 1927), p. 230.

the kind that restructure sentences.

The verb *hit* fits scenes in which something comes into abrupt contact with something else, including those in which some agent manipulates the first of these objects. In the three-element hitting scene, the agent or causer of this event appears as the subject, but apparently either of the other two entities can be realized in the direct object position. That is, we can say either, "I hit the cane against the fence" or "I hit the fence with the cane." (Incidentally, this example demonstrates the need to distinguish what is in the nucleus of a sentence from what are the obligatory constituents of a sentence. *Hit*, it appears, must be accompanied by some mention of the subject against which something moved, even though that object need not be realized as the direct object. In other words, while we can say either, "I hit the fence with the cane" or "I hit the fence," with the same real-world scene in mind, we can say, "I hit the cane against the fence," but not just, "I hit the cane," for that same scene. The contrast between nuclear and peripheral elements is not the same as the contrast between obligatory and optional sentence constituents.)

Using semantic case notions for talking about these sentences, we can say that the things that ended up in the nucleus included the Agent in each of the sentences; some included the Patient, others the Goal; and we say that some representation of the Goal entity is obligatory in the active form of any of these sentences. One intuitive way of thinking about what is going on is to assume that the decision to include things in the sentence nucleus is the decision to take a particular Perspective on a scene. When I spoke of perspective before, I spoke of focusing on a part of a scene, possibly because of some special salience of that part, while being fully aware of the rest of the scene. It is difficult, however, to see how this notion fits the examples we are considering now, since it is not easy to imagine contexts in which we can have the cane in focus but not the fence, and vice versa.

It may be possible to see, at least, what it would mean to let the Agent be out of perspective. Suppose we have a scene in which the Agent swings the cane against the fence. Consider the two sentences: "Harry swung the cane wildly. It hit the fence." Here, even if we understand that Harry was still hanging on to the cane at the time of impact, we have succeeded in putting the Agent's action into the background in the second sentence and are somehow focusing on the contact between the cane and the fence. Here, the two physical objects are in the Nucleus, but the Agent is not.

Let us consider what it would take for us to increase the relative saliency of either the Patient or the Goal in a scene that has the Agent in perspective. This time let us use the word *beat*, which is like *hit* except that it requires the belief that the Agent holds on to the implement throughout the action, whereas *hit* is compatible with a scene in which the Agent lets go. (Consider "Standing on the ground, I hit the third-story window with a brick." Here we get either the scene of a very tall man or of a man who lets go of the brick.)

Since as humans we are more interested in human beings and in human reactions than in what happens to inanimate objects, one way of showing a difference in perspective is to involve another human being in the scene. Consider the sentences "I beat the stick against Harry" and "I beat Harry with the stick." The scene for the first might well be one in which the speaker regards Harry as a nonperson; the second is more compatible with a scene in which Harry says "Oof!" It seems, in other words, that when the goal of a hitting or beating act is a human being, that is, something or someone that is clearly affected by the event, that fact gives the Goal entity sufficient saliency to be included in the Nucleus. If this explanation is correct, it should follow that if the Patient noun designates a human being and the Goal noun does not, the choice of Patient as direct object is the more natural choice. I believe this is true, though with the verb *beat* is a little difficult to separate separate purely linguistic from other kinds of reactions to the test sentences. The point is that it sounds more, let us say, respectful, to say, "I beat Harry against the corner of the building," than to say, "I beat the corner of the building with Harry."

It seems, in short, that one of the conditions favoring inclusion within the nucleus is the condition in which an element is a sentient being who is affected, in the event that is being depicted, in ways we can empathize with. It should follow that if there are two human beings involved that fact alone yields no basis for choosing one or the other. Thus, the two sentences "I hit Harry against Bill" and "I hit Bill with Harry" ought to be and are, I think, equally bizarre.

For situations in which a given noun can be either in the nucleus or in the periphery, one condition favoring inclusion in the nucleus is that of the affected object's undergoing some kind of change. Thus, if I act against something, and if as a result of my action, that something changes, it thereby acquires the saliency to merit inclusion in the nucleus of the sentence.

Suppose, for example, that I am pulling at something and the thing does not budge. In that case I can say that I pulled "at" it or, of course, that I "tried" to pull it, but probably not simply that I "pulled" it. If I pulled at it and it moved in my direction, that is, if it moved closer to me as a result of my pulling at it, then I can say, "I pulled it." Or suppose, to take a similar example, that I am pushing against something and the thing does not budge. I can say that I pushed "against" it, or that I "tried" to push it, but not simply that I "pushed" it. If, however, the thing moves in the direction of my push, as a result of my pushing against it, then I can rightly say, "I pushed it." If you act against something and as a result of that action, the thing moves, that, to repeat myself, gives it the saliency to deserve realization as the direct object.

There seems to be a kind of 'markedness' quality here. If I say, "I pushed the table," you more or less have to believe that I succeeded in getting it to move; if I say, "I pushed against the table," I have not committed myself–except, maybe, in context, by rules of conversational cooperation–to any claims about whether or not the table actually moved.

The verb *break* can be used for a three-entity scene in which someone brings something into contact with something else and *one of those two things breaks*, that is, acquires some discontinuity as a result of this impact. Either the Patient or the Goal noun can appear as the direct object, but only if it stands for the thing that breaks. Suppose, for example, that I swing a hammer against a vase and the vase breaks. I can then say, "I broke the vase with the hammer." As with the *hit* and *beat* sentences earlier, the manipulated object, when it is in the periphery, is marked with the preposition *with*. But now suppose that I swing a hammer against a vase and the *hammer* breaks. This time I must say, "I broke the hammer on the vase." And this time the Goal noun, now out of the nucleus, is marked by a preposition that gives directional information.

The verb *cut* illustrates the same point, though with this verb it is impossible to imagine two things having properties that will allow either one to cut the other; examples with *cut*, in other words, cannot be parallel with those for *break*. Anyway, I can say, "I cut my foot with a rock," regarding my foot as the thing against which a rock was used or manipulated; or I can say, "I cut my foot on the rock," regarding my foot as having acted against the rock. In each case, whether Patient or Goal, the entity that gets cut is the entity that is realized as the direct object of the verb.

Here is a different kind of example. Suppose I take a gun and with it send a bullet toward Harry. I can take the Agent and Source into perspective and say, "I shot the gun"; I can take the Agent and Patient into perspective and say, "I shot the bullet." But if all I did was shoot the bullet in Harry's direction, then I cannot bring Harry into perspective. In such a case I can say, "I shot at Harry," or "I shot the gun at Harry," or "I shot the bullet at Harry"; but only if I succeed in hitting Harry with the bullet can I say, "I shot Harry." That is, only the scene that includes hitting Harry as a consequence of the shooting act is one in which Harry has the saliency to be included in the nucleus of the sentence.

This is the place to make clear that this speculation about saliency and the nature of the development of scenes is not in itself an explanation of, for example, why *Harry* can be a direct object in "I shot Harry," but rather an attempt to account for the conditions under which a verb is allowed to take a Goal noun as its direct object. There is still the lexically specific information to contend with. A verb very similar to *shoot* is *fire*, but *fire* does not have the same set of options with the Goal noun phrase. We can say, "I fired a gun," or "I fired a bullet," or "I fired at Harry," but not, with the same scene in mind, "I fired Harry."

Sometimes a Source, Goal, or Range constituent completes the sentence in such a way that the action described by the sentence acquires in total some special salience that allows the associated nonsubject noun- phrase to become the direct object. If it were no particular accomplishment to swim across the English Channel, we would only speak of the deed as "swimming across the Channel" not as "swimming the Channel." If leaping or jumping across something is not taken as any kind of special feat, we will speak of leaping across the thing; if the deed takes on some special salience, by virtue, say, of the fact that the thing across which one leaps is an obstacle, then we speak of leaping the thing. Thus, as Jeffrey Gruber (1965, p. 24) has pointed out, we can talk about either "leaping across the wall" or "leaping the wall," since leaping across a wall can be viewed as an accomplishment of some importance; but while we can speak of "leaping across a line" (drawn on the sidewalk), it is hard to imagine a scene appropriate to the sentence "He leaped the line."

I mentioned earlier that the verb *sign* can take as its direct object either a noun designating the product of the writing act–as in, "He signed his name on the contract"–or a noun that designates something that acquires some new importance by virtue of the signing act–as in, "He signed the contract" or "He signed the check." Here again the acquired validity of the thing gives it the needed saliency. Notice that although we can say, "He signed his name on the men's room wall," it would be odd to say, "He signed the men's room wall."

Let me turn now to the examples Anderson used in his arguments in support of the standard theory's level of deep structure. They have to do with familiar pairs like, "I loaded the truck with hay" as compared with "I loaded hay onto the truck," and "I sprayed paint on the wall" as compared with "I sprayed the wall with paint." There is a natural or unmarked choice of direct object–namely the Patient noun–only when there is nothing special to say about the Goal. However, if the action with respect to the Goal acquires saliency by virtue of being in some sense 'complete,' then the Goal noun comes into the nucleus. Thus, for these sentences, the act of "loading the truck with hay" is most naturally seen as "filling" the truck with hay, and the act of "spraying the wall with paint" is most naturally seen as "covering" the wall with paint. This is Anderson's holistic interpretation. Anderson's way of saying it is that we start out with subjects and objects defined by phrase structure configurations, and the semantic component then assigns a holistic interpretation to certain nouns, in certain contexts, only when they are in subject or object position. My way of saying it is that a verb is sometimes capable of having its subject/object structuring accomplished in more than one way, and that such saliency properties as totality or completion justify bringing certain nouns, depending on their underlying case role, into the nuclear structure.

It should follow, then, that there can be verbs matching scenes that have the totality or holistic interpretation built in. This is true, for example, of the verbs *cover* and *fill*, the verbs I used in explaining the saliency conditions mentioned in the last paragraph. Thus, "He covered the alarm clock with the towel" is English, while "*He covered the towel over the alarm clock" is not; and "He filled the jar with ink" is acceptable, while "*He filled ink into the jar" is not. By contrast, the verbs *place* and *pour* require focus on the object-manipulating rather than the resulting-state parts of their scenes. They allow expressions like "He placed the records on the shelf" but not "*He placed the shelf with records," and "He poured ink into the jar" but not "*He poured the jar with ink."

It should be emphasized that idiosyncratic lexical properties are at play here too. There is no noncircular way of describing a verb as necessarily focusing on, say, the Agent-manipulating-Patient part of a scene. Judging from the meaning of the verb *hang*, one would expect that it might require such a focus; yet it is possible to say things like "He hung the wall with drapes." And in Hungarian, which allows many of the same valency changes as I have been illustrating for English, the verb *tölt* translates as both "pour" with Patient as direct object, and English "fill" or "stuff" with Goal as direct object.

It seems, in short, that the decision to include nouns in the nucleus of the sentence or to leave them in periphery is guided, at least in part, not so much by any understanding of the action itself, but by some notion of the saliency of the action as a whole or the saliency or importance of the specific entities involved in the scene that make one portion of the scene specially deserving of focus. I feel sure that there must be some random historical forces at work here too, but I frankly do not know of any completely noncircular way of showing what I mean. (David Justice has brought to my attention Baugh's report (Baugh 1957, p. 387)that in South African English the verb *throw* can take as direct object a person hit by the thrown missile; thus it is possible to say, "I threw Harry over the hedge with a rock." Niels Ege has told me that the

Danish verb meaning "shoot" takes an animate Goal as direct object only if the result of the shooting is that the being that was hit actually dies. And I have found in the current class of students in Berkeley that there is a fairly common use of the verb *win* with a human direct object referring to the opponent in a competition. These young people can say not only "I won the game" and "I won the prize," but also "I won Harry," meaning "I defeated Harry.")

6. The second problem for the mapping from scenes onto underlying structure was that of determining the ranking of elements put in the nucleus. For this I suggest that there might be a Saliency Hierarchy that serves several purposes in a grammar. In the first place, since every sentence has to have a subject, the scene entity that has the highest rank will be realized as the subject; in the second place, if two entities are in the nucleus, that is, are in perspective, the roles of the first and second terms will be allocated according to relative position in the hierarchy. In the third place, the hierarchy imposes a constraint on what can appear in the nucleus and what in the periphery for the same verb. If it is a verb that can take either of two things as direct object, the one that outranks the other on the Saliency Hierarchy wins out.

At present this is sheer speculation.[5] What I have in mind is something to replace the Case Hierarchy of my work on case grammar. Now what I say is that we have a scene, with some part of it put in perspective. A saliency hierarchy that provides a mapping principle between this perspective and the structure of the sentence in terms of grammatical relations will look something like this:

1. An active element outranks an inactive element.
2. A causal element outranks a noncausal element.
3. A human (or animate) experiencer outranks other elements.
4. A changed element outranks a nonchanged element.
5. A complete or individuated element outranks a part of an element.

5. The way of dealing with the "hierarchy" presented in these lectures differs somewhat from what I suggested in another paper written at about the same time, "The case for case reopened," (this volume, Chapter 5). The view I present in that paper is more or less like the "case hierarchy" I proposed in Fillmore (1968): that is, noun-phrase roles were determined by their fit into the preestablished hierarchy of role types, their relative ranking determined by the ranking provided by that hierarchy. Here, however, I am suggesting that such matters as subject-selection (of the "unmarked" kind) might be determined by an ordered set of ranking considerations. I do not have good reasons for preferring one of these proposals to the other. The main advantage I see to the proposals offered in this lecture is that it eliminates the need to determine the "final, correct" list of cases.

6. A 'figure' outranks a 'ground.'
7. A 'definite' element outranks an 'indefinite' element.

The intention is that this hierarchy is to be consulted in the order in which these statements are listed. Thus, an active element outranks everything else; a causal element outranks everything but an active element; and so on.

If two elements in a scene are of equal rank by the Saliency Hierarchy, then it is possible to put either one of them into perspective. Thus, in a commercial act, it is possible to take either the buyer or the seller into perspective, but only by regarding the other as the Source or Goal, rather than as an Agent. In locating something with respect to something else, it is possible to take either one as the figure and the other as the ground. Thus, we can speak of the map on the wall being under the picture, or the picture being over the map. The actual physical scene does not determine this choice, only the speaker's need to take one of the objects as the reference object.

7. Let me turn now to the commercial event, and make some initial proposals on a notation for scenes, perspectives, and grammatical relations. The elements in a prototypic commercial event scene are the buyer, the seller, the money that changes hands, and the goods that change hands. Let us regard as the prototype scene for the commercial act one in which a seller sells what he owns, rather than, as a salesman, what somebody else owns; that the buyer buys by rendering money, not services or favors; that what is exchanged is goods, not services; and that the transaction occurs as a single, completed event, with no IOUs or down payments.

Let us use as an example a scene in which our friend Harry buys a puppy for $60. Here are some of the things we can say about this event.

Harry bought the puppy.
Harry bought the puppy from Mr. Smith.
Harry bought the puppy for $60.
Harry bought the puppy from Mr. Smith for $60.
Harry bought the puppy with the $60 that his mother gave him.
Mr. Smith sold the puppy.
Mr. Smith sold the puppy to Harry.
Mr. Smith sold the puppy for $60.

Mr. Smith sold the puppy to Harry for \$60.
Mr. Smith sold Harry the puppy for \$60.
Mr. Smith sold Harry the puppy.
Harry spent \$60 on the puppy.
Harry spent \$60 for the puppy.
Mr. Smith charged \$60 for the puppy.
Mr. Smith charged Harry \$60 for the puppy.
Mr. Smith charged Harry for the puppy. (He didn't give it to him.)
The puppy cost \$60.
The puppy cost Harry \$60.
Harry paid for the puppy. (He didn't get it free.)
Harry paid Mr. Smith for the puppy.
Harry paid \$60 for the puppy.
Harry paid Mr. Smith \$60 for the puppy.
Mr. Smith priced the puppy at \$60.

Let Fig. 1 stand for the commercial event as a whole; and let the strip in Fig. 2 represent an analysis or unpacking of Fig. 1. In Fig. 2 the left-to-right order of the boxes indicates change in time, and vertical stacking indicates simultaneity. A and D are the buyer and the seller, respectively; B is the goods; C is the money. The figure is intended to

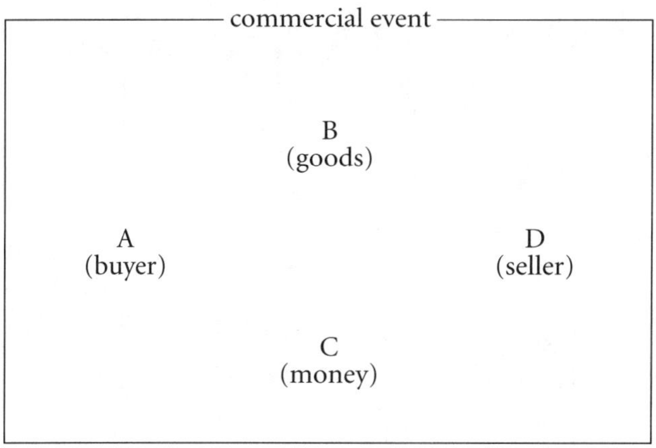

Fig. 1

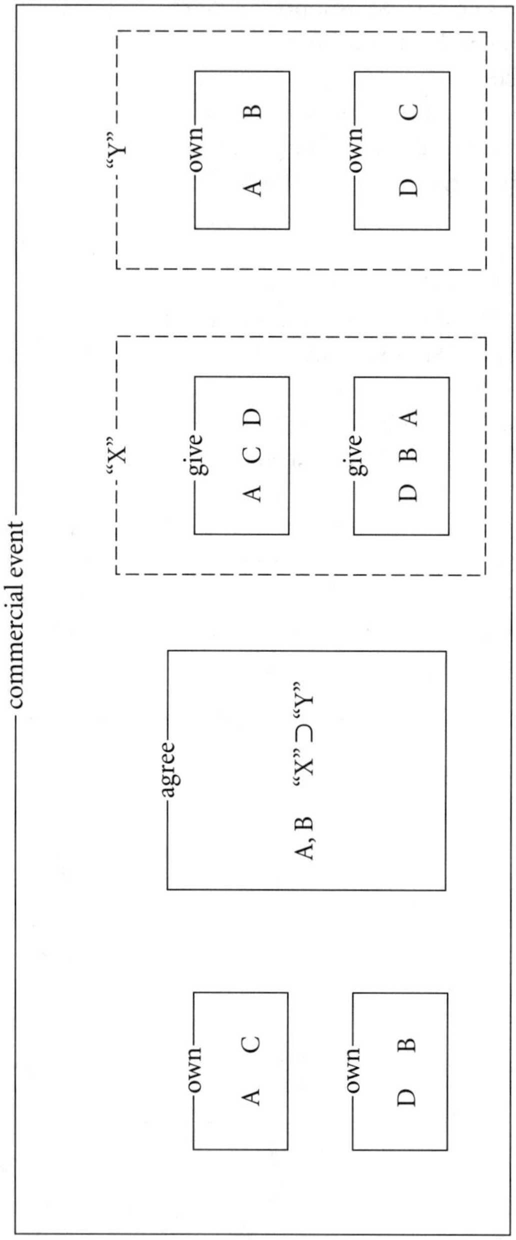
commercial event
own
A C
own
D B
agree
A, B "X" ⊃ "Y"
"X"
give
A C D
give
D B A
"Y"
own
A B
own
D C

Fig. 2

show that in one time slice, A owns a certain amount of money C (he may own more than that too, of course), D owns B, and it is agreed between A and B that an exchange between them of these things will result in a transfer of ownership. In the next time slice A gives D the money and D gives B, whatever that is, to A. In the final time slice, A owns B and D owns the money.

Since there is nothing in Figs. 1 or 2 indicating perspective, they represent a scene schema rather than the meaning of some linguistic form. The labels on the boxes indicate scenes that either have to be simply known by experience or need to be explained or defined by more detailed diagrams of the sort given here.

With the larger diagram in Fig. 2 taken as a kind of definition of the commercial act, we can now construe Fig. 1 as a schema for a commercial act taken as a single situation, and identify the participants as A, B, C, D, according to their roles in the scenes specified in the definition. In the diagrams given below as Figs. 3 through 9, the part of the scene that is put in perspective is in an enclosed curve; the grammatical relations are indicated with the numbers 1 and 2; peripheral elements that are expressible as complements to the verb are marked with their appropriate prepositions; and elements from the scene that cannot be included in a simple clause whose main verb is the label of the box are put in square brackets. A Goal element that is obligatorily made into an indirect object is marked with the number 3.

8. What I am after, of course, is a representation that is informal and easy to work with, so that lexical differences can be displayed in a fairly straightforward fashion. The complications are many, but taking this as a goal is one way of discovering just what the complications are. They include the necessity of defining the larger scene (and that includes being clear about the temporal relationships); the difference between obligatory and optional elements; the difference between nuclear and nonnuclear elements; the grammatical relations of the elements in the nucleus; the grammatical markings of the elements in the periphery; and–something that is disguised here–the aspects of other scenes that must also be integrated with these scenes, or ways in which the lexical item being displayed here refers to a larger scene than the one being discussed.

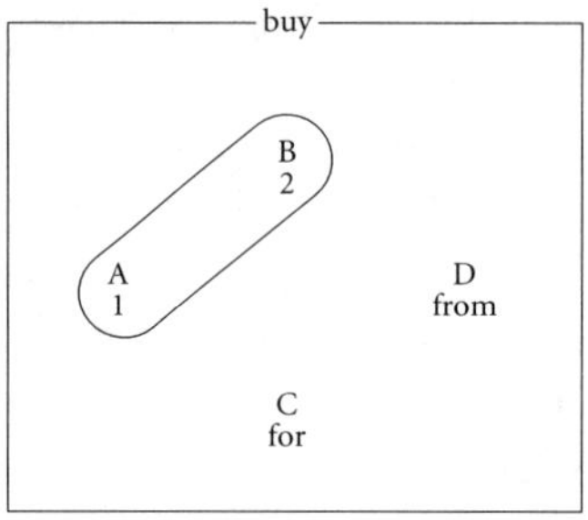

Fig. 3

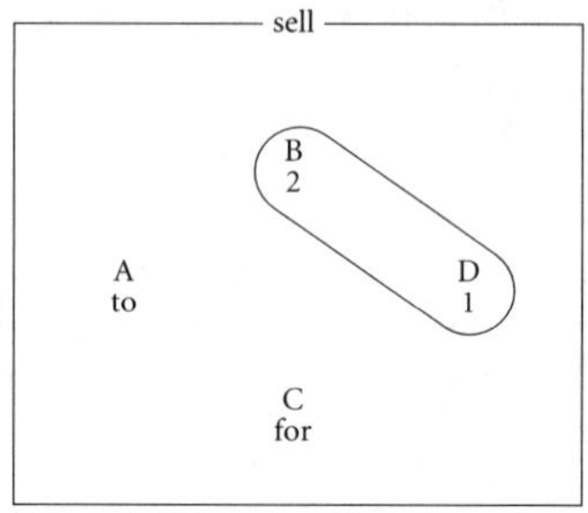

Fig. 4

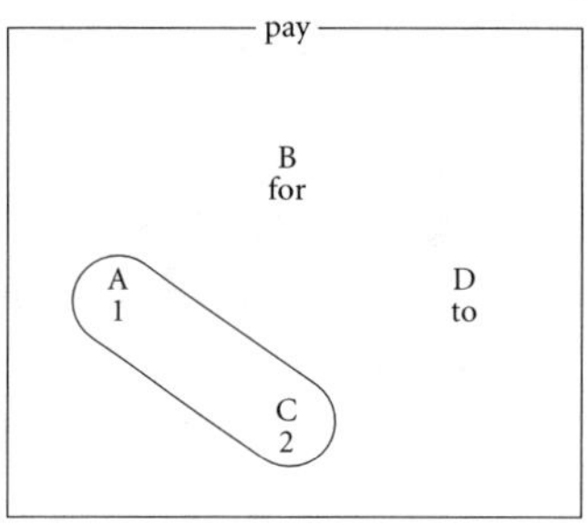

Fig. 5

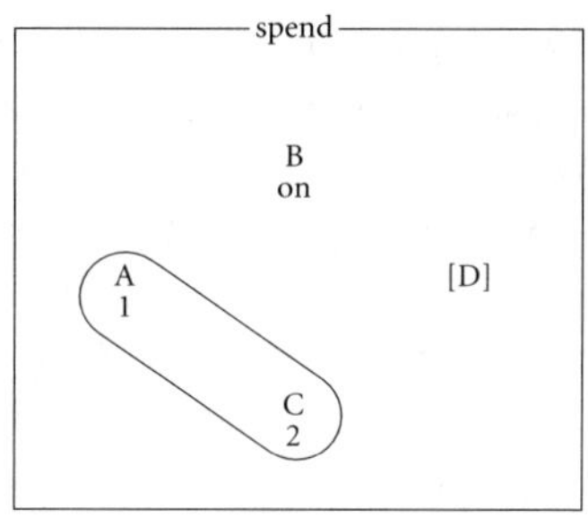

Fig. 6

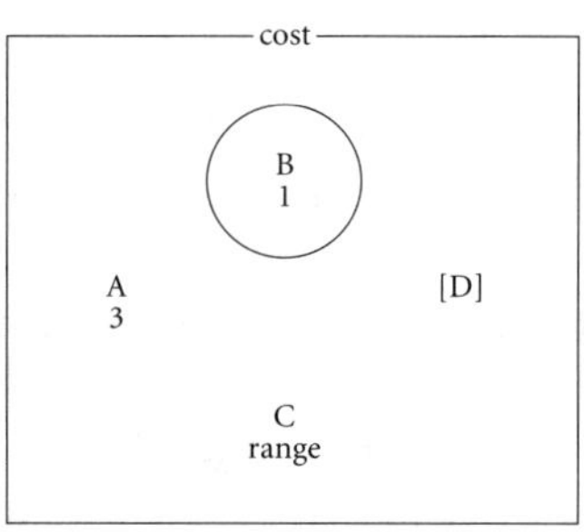

Fig. 7

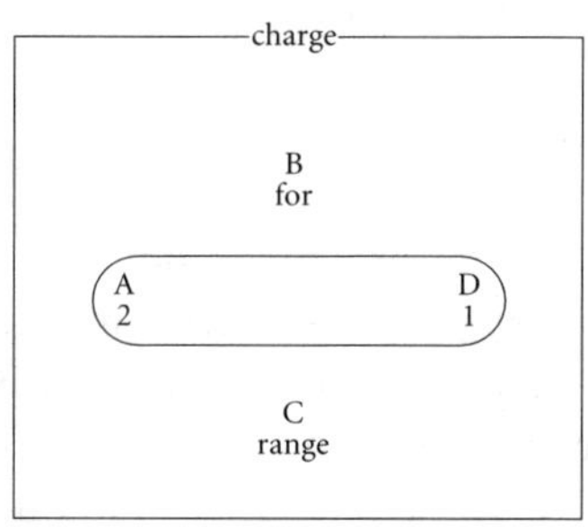

Fig. 8

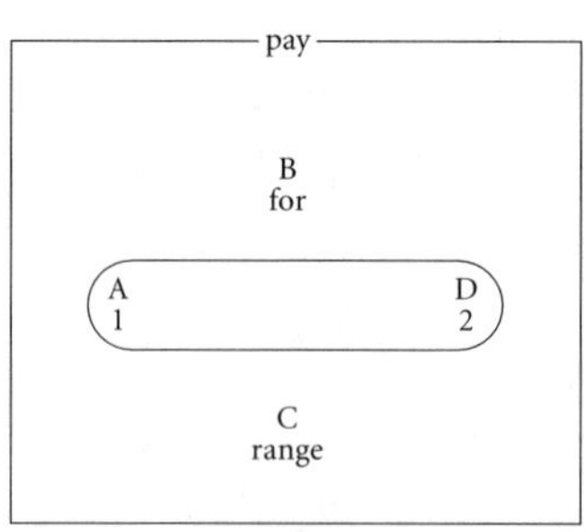

Fig. 9

A detailed description of these verbs will have to point out that with *buying*, prepositions distinguish reference to the actual cash and reference to the value of the money; thus, we can say, "I bought it with the silver dollar I found in the attic," or "I bought it for a dollar." With *selling*, on the other hand, the preposition *for* is used in both cases: "I sold it for a dollar" and "I sold it for this silver dollar." *Spending* can name a relationship between a person and money over a series of purchases, as in, "I spent seventy dollars last week," as well as the relationship between a person and the money spent in a single commercial event. *Paying* can involve any prearranged required transfer of money, as in paying debts, paying a child his allowance, etc., and is not limited to commercial event scenes. *Charging* can be used without any notion of an exact money amount, indicating simply a contrast with giving something for nothing, as in, "He usually gives it to me free, but today he charged me for it." *Costing* can refer to a static relation between an object for sale and its price, independently of any actual or specific buying/selling act. This can be seen in a sentence like, "If it didn't cost so much money, maybe somebody would buy it."

By taking a kind of simple prototype commercial event, we can examine, one by one, the lexical items that activate or are activated by scenes of such an event. By comparing, intuitively, what we know of the use of these words with the properties of the scene as we have set it up, we have a framework for discussing the words that require more information–more complicated scenes–as well as those that are actually more general, that require less detailed scenes, than the prototype we set up. I have only scratched the surface with this collection of verbs. In the next lecture I will explore money giving with a little more detail, I will suggest some more possibilities with the notation, and I will end by saying something about how cognitive scenes of the type we have been considering can be seen to integrate with scenes found in the current context of interaction in a conversation.

LECTURE THREE

1. In my second lecture I proposed a simple method of diagramming certain properties of what I referred to as the 'scenes' that 'activate' or 'are activated by' given linguistic expressions. I intended these to represent aspects of what people have in mind when they comprehend texts containing the expressions.

My hope was that this kind of diagramming will make it possible to present in an intuitively graspable way simple aspects of word meaning and text meaning. It remains to be seen whether or not the result will still be satisfying after we have devised the abbreviations and conventions that will surely be necessary for displaying subtler aspects of meaning than those I will be illustrating. In any case, I believe that by exploring this method for what it is worth, we can gain some clarity on the nature of some of these subtler factors. In that sense our notation can serve as a kind of 'discovery procedure' for semantic exploration.

Where time development is relevant in characterizing a scene, the notation has some resemblances to a comic strip. Thus our method begins by representing in a fairly simple way something that would require in most other notations cumbersome formulations of the type

> Event *e* occurred at or during time *t* and event *e′* occurred at or during time *t′*, and *t* is earlier than *t′*.

In addition to mere sequence in time, of course, we need to know what aspect of a complex scene is being focused on by the speaker through his use of a particular linguistic form at a particular point in the text. For this we can add an arrow to our comic strip notation. Recall that in order to understand our earlier sentence about the hospitalized Mark *sitting up* we needed to have in mind an earlier state during which Mark was *lying down*. Thus, using the arrow for the focus, we can represent *sitting up* in a way suggested by Fig. 10. The sentence is saying something about the state pointed to by the arrow; but the person who understands that sentence understands that the state being pointed to is a stage in a history; and he knows something about the nature of that history–the rest of the strip represents that knowledge.

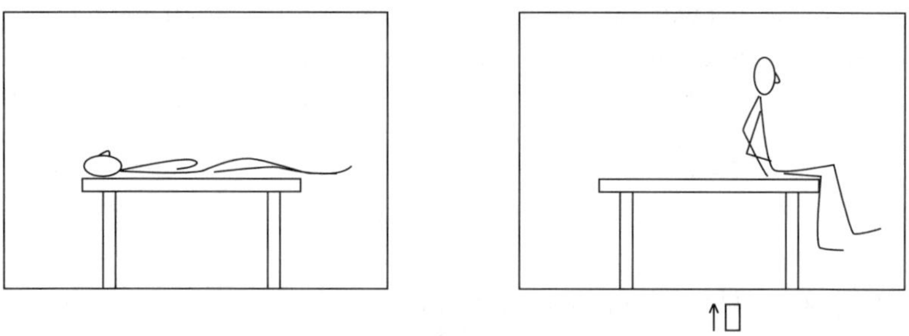

Fig. 10

A second property we need to be able to represent for natural-language semantics is simultaneity. If left-to-right order of scenes is used for showing temporal sequence, then the most natural way to indicate simultaneity is vertical stacking. This was something we needed for talking about the various stages in the commercial event. The resulting state in that event, for example, was one in which the seller owns the money and the buyer owns the goods. That state of affairs is indicated by the stacked boxes given here again as Fig. 11. Comic strips, of course, do not have anything corresponding to our vertical stacking. Instead of thinking of the separate boxes in our notation as representing individual frames in a comic strip, we can think of them as providing instructions to an artist concerning the objects and the relationships that ought to be depicted in the total scene that occupies that part in the strip. For our example, the artist merely needs to know that the scene must show A as the owner of B and D as the owner of C. As the amount of stacked information increases, however, as is the case in the development of a large text, the resulting picture becomes more and more Breughelian.

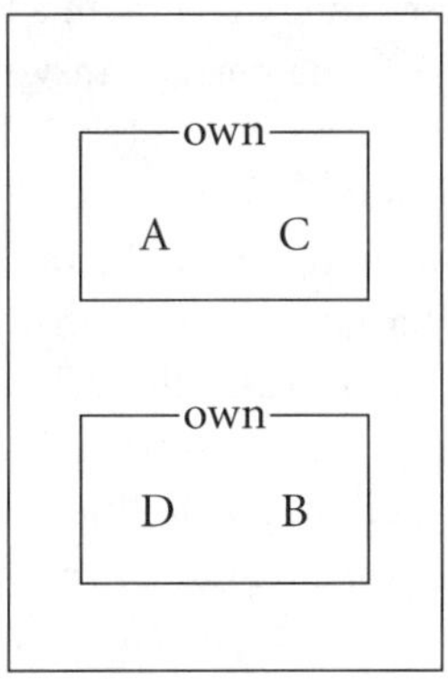

Fig. 11

I hope, by the way, that my appeals to visual imagery are more helpful than misleading. I obviously do not mean what I say literally, so that is why, instead of trying to draw a picture of somebody owning something–that, after all, is impossible–I merely provide each scene or subscene with a label that identifies the property or situation or relationship, listing variables standing for the entities in an order dictated by the saliency hierarchy.

2. In addition to sequence and simultaneity, a third aspect of scenes that we need to represent I will call "selection." I have not had anything to say about it so far. Sometimes a meaning is relative to a scene that offers a selection from two or more possibilities; and sometimes individual words are understood as identifying one choice within just such a contrast set. The meaning of the

word requires, then, an understanding of the nature of the contrast within this particular scene.

For example, there is the scene connected with the kind of commercial event that takes place in a store. At least two situations provide for the use of the word *cash.* In one of these, the contrast is between paying in currency as opposed to paying by some other means. In this setting, the word *cash* rules out the possibility of payment by personal check. In another situation the term opposed to *cash* is *credit.* Within this setting the choice of credit requires the clerk to write up the sale on a special form, to ask for the customer's signature, and perhaps to check the customer's credit rating; here the word *cash* can be taken as including the possibility of payment by check. By knowing something about the nature of these two contrast sets, we can know that the corner tobacconist who does business for cash only will not accept a check; and at the same time we know that if a clerk in a large department store asks, "Will that be cash or charge?" the correct answer, if we intend to pay by a check is "cash."

In each case the important difference is between what is particularly salient or troublesome, and its opposite. In the one setting, payment with currency is safer than the alternative; in the other, payment with a credit card is the more troublesome. In both cases these 'marking' properties are determined from the merchant's or salesperson's point of view. The word *cash,* in other words, designates different possibilities for payment depending on which contrast set it belongs to. One way of representing this state of affairs, in the cash/charge contrast case, is by means of a diagram like that shown in Fig. 12, which is to be presented in a simultaneity relationship to a commercial event. We can insert a variable representing the commercial event in the appropriate cell.

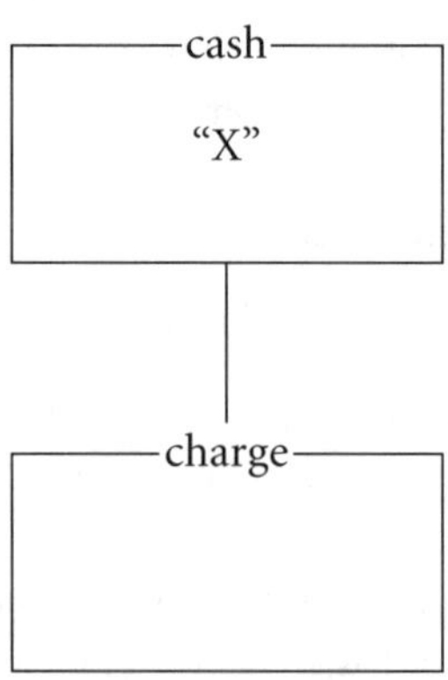

Fig. 12

3. Some scenes involve understandings about contracts, conventions, laws, expectations, etc., and these may require cross-references linking one part of a scene to another. To indicate these cross references, I propose using the late letters in the alphabet (W, X, Y, Z) as variables, in quotes, both within the statement about the agreement (or whatever) and as indicators of the portions of the scenes to which they refer.

For example, something counts as a commercial event, not just by virtue of an exchange of a sum of money and some goods resulting in ownership changes; that situation, after all, is compatible with a simple coincidence of two gift-giving acts. There has to be an agreement, in advance of the act, on the part of both parties, that the exchange will in fact result in the ownership changes.

A part of the notation for the commercial event presented in the last lecture is given here again as Fig. 13, which indicates that at one stage in this history, A and D agreed that after the event labeled "X" occurs, the state labeled "Y" will necessarily obtain.

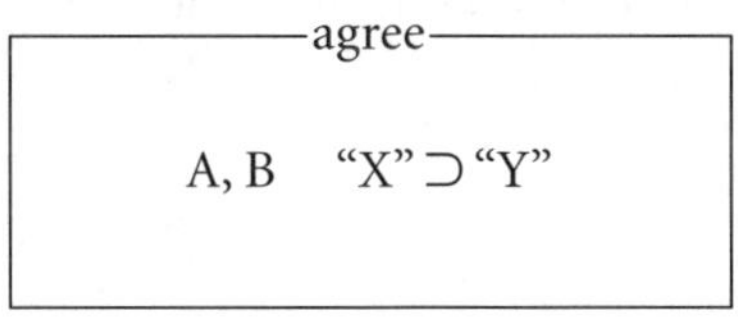

Fig. 13

Here, too, it is necessary to repeat that what I have in mind when I talk about 'the commercial event' is a prototype instance of such an event. A sign in a china shop that reads, "Break it and you've bought it," identifies a situation that departs a little from the prototype. Most cases of buying and selling, in fact, do not involve explicit agreements made by the two participants in the transaction but rather agreements that are implicit in the situation. If there is a price tag on the bar of soap you are about to buy, there is no reason, in most parts of the English-speaking world, at any rate, for you to make any on-the-spot agreements with the clerk on the amount of money he requires in exchange for the soap. (The terms of the prototype I have set up are perhaps more directly appropriate for primitive markets than for supermarkets.)

4. Another aspect of the diagram for the commercial act was the presentation of conditions that have no temporal position in the scene at all, but that simply hold throughout. Information of that kind includes what we might think of as the 'cast of characters' and general descriptions of the larger setting for

our scene. For the representation we devised for the commercial event this was the information that C was money, B was goods, and A and D were people. We can put such information in a box that stretches through the whole strip.

5. So far I have spoken mainly of scenes associated with verbs. Remaining within the money realm, I would like now to examine an application of this notation for describing the meanings of the various nouns that speakers of English use when talking about sums of money exchanged in specific money-transferring scenes. Suppose we know that on a particular occasion, one person gave another person some money. Consider what must be assumed about the larger scene of which this scene is a part, once we know what the money, on this occasion, was called. A metaphor that fits the view of semantics I am proposing is that when you pick up a word, you drag along with it a whole scene. The insight beyond that metaphor might become particularly clear if you try to imagine, for each of the following nouns, something about the nature of the larger scene. Consider them one at a time.

tip	bonus
ransom	rent
allowance	fare
refund	child support
honorarium	bus money
bounty	salary
tuition	reward
retainer	alimony

The list could go on, of course.

Any semantic notation needs to account for the scenes introduced by nouns, too. More often than not, the richness and detail of the large scene we eventually construct for a text may be more dependent on scenes introduced by nouns than on those introduced by verbs.

Consider for a moment the word *alimony.* If A pays alimony to B, we know that at one time A and and B were married, we know that the marriage ended in divorce, and we know that roughly at the time of the divorce an agreement was made between the two participants to the effect that one of them would pay money to the other. The prototype, which these days may be losing its prototype, has it that the ex-husband is the one who pays money to the ex-wife.

We know that this history must have included more than that, too, so that if we were to dwell on this example to fill in more of the details, we might have images of bitter disagreements, lawyers, court appearances, court fees, public notice, and a host of other things.

The part of this larger scene that seems to be necessarily assumed whenever some exchange of money is correctly referred to as the payment of alimony is represented here as Fig. 14. The diagram is a representation of the meaning of the word *alimony*: we see in the diagram that the lexical item *alimony* is linked by an arrow to the entity C, a fixed sum of money, in the rightmost box. Other aspects of the total scene we were able to imagine are associated with the concept *alimony* only indirectly, by means of prototype scenes associated with legal acts and with the marriage relationship.

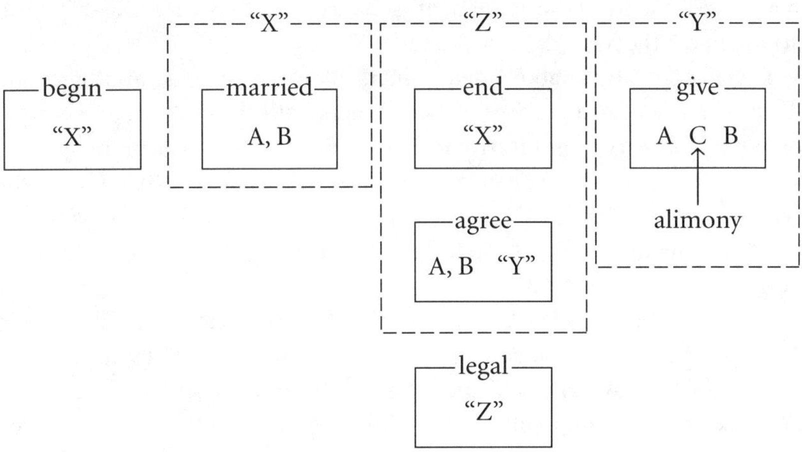

Fig. 14

As a second example, let us take the noun *ransom*. If what A pays B is called RANSOM, we have the right to assume that B had illegally taken possession of some person or object C that is important to A, that B had offered to return C on condition that A give B some money, and that A had accepted this offer. A comic-strip representation of *ransom* would include the information that at one stage B took illegally something valued by A; at a later stage B offered to return what he took on condition that A would agree to give B a certain sum of money; at a third stage A gives B the money, which is then properly called *ransom*, and B takes on the obligation to restore what he took.

6. In my discussion of nouns like *alimony* and *ransom* I have so far been talking as if what we were dealing with was merely a matter of choosing a noun

for referring to some amount of money that changed hands. These examples allow me to make another point about scenes and texts. The expressions that we need to embed these nouns in are not uniform. Some situations call for an indefinite determiner some for the definite determiner and some for a possessive pronoun and some allow the verb *pay*, others only *give*.

If it is clear that A is expected or obliged to give the money to B, the possessive pronoun is appropriate. Thus: "I gave him his allowance, his salary, his change." If the presentation is not expected or obligatory, the indefinite article can be used. Thus: "I gave him a tip, a bonus, a reward." Sometimes before a money-exchange agreement is made, one can talk about it with the indefinite article; once the agreement is complete, a possessive or definite article might be necessary. Thus I can say in advance, "I'll give you an honorarium, a salary, an advance," but after the agreement is made, I must say, "Here's your honorarium, here's the reward," or whatever.

These observations about determiners allow me to go to another point. The possibility of using either of two choices with the same noun shows us something of the communicative force of using a linguistic form that presupposes some specific scene or expectational set. If a waiter tells me, "You didn't give me a tip," I get the feeling that he finds me inconsiderate; if he says, "You didn't give me my tip," I get the feeling that he is telling me that I did something illegal.

What I am suggesting, of course, is that the semantics of the determiner system interacts with the semantics of the scene activated by a particular noun, in such a way that the choice of one or another article, or the choice of the possessive pronoun, will contribute in important ways to the construction of the final completed scene. One of the disadvantages of scene semantics, in short, is that it does not allow you, without embarrassment, the convenience of talking about the semantics of verbs and nouns while leaving to the future the problem of incorporating into your system information about the semantic effects of articles. Some choice from the determiner system must accompany any noun, and the scene the interpreter constructs depends in many ways on just what choice was made.

7. Still another lesson to be drawn from these examples concerns the ways in which scene schemata associated in the first instance with an individual lexical item can play a framing role in the assembly of the scene representing the meaning of a whole sentence or text. We are not going to encounter the word *alimony* only in word lists; we will find it embedded in a sentence. For example, in the sentence "Harry paid up his alimony," the scene associated with

alimony provides the framework for the scene that matches the whole sentence. We identify the husband in the alimony story as Harry, and we add the information that the financial agreement was fulfilled, that is, that the obligation has ended, with an arrow pointing to the box that expresses that.

But now let us illustrate that same point with a more complicated sentence, namely, "Harry paid this month's alimony with money borrowed from his stepdaughter." An analysis of this sentence will be my main demonstration of the connection between lexical semantics and text semantics.

The complex scene that we construct out of this sentence will have the marriage-divorce contract-alimony payment sequence that we have seen in association with the word *alimony.* The phrase "this month's" adds the understanding that the payments were, presumably as part of the contract agreement, to be paid in monthly installments. The noun *stepdaughter* introduces a more complicated history to Harry's life. In order for him to have a stepdaughter he has to be married. For him to have a stepdaughter requires the understanding that his current wife was previously married and had at least one child by that marriage, and that that marriage ended. For Harry to have borrowed money from his stepdaughter means that he has entered into an agreement by which the money she gave him on this occasion must be returned at some later date. My point is, of course, that in our effort to construct a single history to fit the whole sentence, we have to fit into each other, in some satisfactory way, the various relationships and time lines that the separate parts of the sentence have introduced.

The example allows me to show something of the separate roles of linguistic and extralinguistic knowledge in an interpreter's processing of a text. We know that Harry used to be married to one person and that he is now married to another. We know these things because he is described as paying alimony and as having a stepdaughter. Our knowledge of what the world is like–at least the familiar world–includes the knowledge that a man is married to at most one woman at a time, and requires us to add to our text scene the assumption that Harry entered into his current marriage after the time of the divorce we have been considering.

There are still temporal relations that we do not know, however. This sentence, in fact, gives us a new reason for being dissatisfied with the comic-strip notation. We cannot know anything for sure about the temporal ordering of, say, the beginning of Harry's first marriage and the birth of Harry's current wife's daughter or the beginning of Harry's current wife's former marriage. It seems that what we need is a modified kind of comic strip with several tracks, printed on rubber sheets. The two sections corresponding to the two separate histories are attached at the point where their histories are known to coincide, but their beginnings are left hanging, with temporal relations between them unspecified. If later on in the same text we learn that Harry's first wife was born on the day his second wife's divorce became final, then we can stretch out the two rubber strips and staple them together at the point where the text shows a temporal coincidence of two points in these two separate histories.

The process of interpreting a text involves beginning with some sort of scene, letting that provide a framework for embedding and overlapping and connecting with other scenes. I have emphasized that the way in which this is done involves meanings of words and meanings associated with particular grammatical choices, and much more. When we interpret a text, we bring to the task more than our knowledge of the language–knowledge about the world, beliefs about human nature, assumptions about typical instances of objects, repertories of stereotypic instances of behavior, and so on. All of this contributes to the ultimate picture we get of the world that matches the text.

8. Let us return once more to Mark and Mike in the hospital. Because of the linguistic forms *sitting up* and *sitting down* we were forced to make assumptions about their previous body positions. But there were certain other assumptions that we might have found ourselves making, too, and these had to do with our beliefs about why the boys might be in the hospital, what sorts of changes in their condition would have induced the nurse to make the remarks that she made, and so on. These details we found ourselves constructing are not strictly authorized by the text, but are added by the interpreter. Sometimes we are more aware of the part that we added than at other times; and if people's experiences are different, it follows that people differ in the sense they make out of the same text.

Highly trained and sensitive individuals, such as the ideal jurist or the ideal psychiatrist, can probably always keep in mind the difference between what a text actually authorizes and what they as interpreters have added to it. Schizophrenics, I would guess, are the least able to do this. They can believe that they have communicated to others, or that others have communicated to them, details that could only have been inferred, from what was said, by somebody with a very special set of beliefs and a very special kind of imagination. Most of us fall somewhere between these two extremes.

I have suggested that the interpreter of a text contributes to the interpretation process in ways that are independent of the meanings authorized by the text itself. There are, of course, more basic ways in which the participants in a conversation are involved in a text; and here we enter the realms of pragmatics, sociolinguistics, and deixis. Sometimes the linguistic forms found in a text themselves anchor the scenes that match the text to the world containing the conversation participants themselves. The comic strip needs to include a scene in which somebody is saying precisely the sentence that is being analyzed. Actually, the sentence we were just considering is like that to a certain extent. We have to believe that the utterance of the sentence occurred during the month that the speaker refers to in the expression "this month's alimony check."

9. The kind of analysis I have been demonstrating is text interpretation from the overhearer's point of view. In a slight fear of being misunderstood, we might think of what I have been discussing as an effort toward a 'competence' theory of discourse semantics. The task is to determine what we can know about the meaning and context of an utterance given only the knowledge that the utterance has occurred. If we know, for example, that somebody has used the sentence "You may come down out of the tree now, sir," there are many conclusions we can draw about what is being communicated, how the communicators are related to each other, socially and in physical space, and so on.

Many people regard this kind of research as evil because it ignores real-world contexts. I think the only superiority of real contexts over constructed ones is that they sometimes make you realize things you would never have thought of if you had stayed forever alone in your rocking chair. I am interested in language use, too, but I find that whenever I notice some sentence in context, I immediately find myself asking what the effect would have been if the context had been slightly different, if a different word had been used, if the sentence had been given a different intonation, and so on. In other words, I get right back to a 'competence' theory approach.

The goal in doing discourse analysis from the overhearer's point of view is that of finding the lowest common denominator among the set of possible associated scenes and contexts. Any actual use of a sentence will have some or all of the background conditions satisfied, will have some or all of the unspecified choices specified, and so on. The only uniform description that can be given to texts independently of contexts is one that characterizes in the most general way the *set* of contexts in which it could be put to use.

10. From the point of view of the interpretation process, pragmatic knowledge can be thought of as the knowledge by which we are able to construct a scene of the setting in which the text was produced—that is, to contextualize it. If the text happens to be a conversation, then this pragmatic knowledge

enables us to understand something of the setting, the conversation partners, the locations, and maybe the postures and movements of the participants.

To show the contribution of pragmatic knowledge to the scene-construction process let us consider the simple sentence "Mommy will come there tomorrow morning." The scene we construct for a sentence that contains deictic elements has to contain direct reference to the utterance being analyzed. We manage this by including in the scenes we construct the event of somebody speaking the utterance being examined.

We begin by locating the event of somebody saying "U" (="Mommy will come there tomorrow morning") within a calendar day. Since anything at all occurs within a calendar day, that is not so significant; but we do this because we will need to refer to the next calendar day in the sequence. Fig. 15 is a representation of such a scene, in which A is the speaker of "U," and B is the addressee. The time line is represented by the segmented arrow at the top; the divisions in the time line are divisions between successive calendar days.

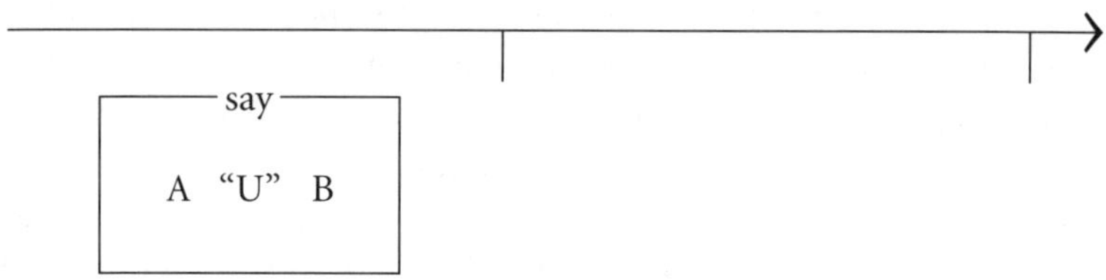

Fig. 15

Our sentence contains the word *come*, referring to a movement from one place to another. We could represent the movement part of this meaning as Fig. 16, in which D is to be taken as the thing that moves, E as the starting point, and F as the destination of the movement. As with the analysis of the commercial event, it would be possible to break up the movement scene into a sequence of temporally ordered subscenes representing D's location at E at one stage and D's location at F at a later stage.

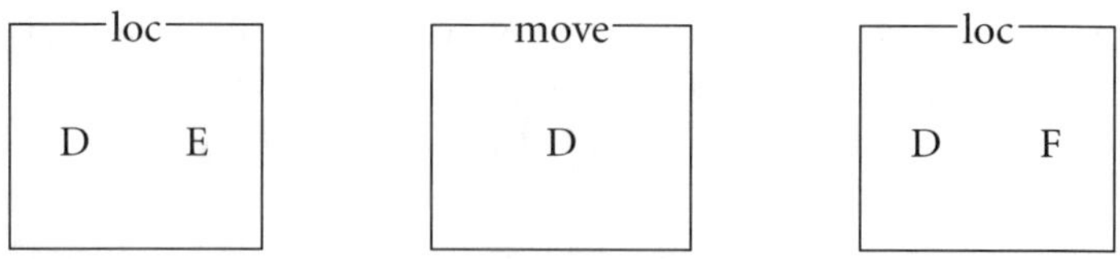

Fig. 16

The word *Mommy* is used in this text. Mommy happens to be a word used by children and by adults when speaking to children. Thus we conclude that either the speaker or the addressee is a small child, or that both of them are.

We can say, taking the competence view I have been suggesting, that the text at this point is compatible with three different scenes, depending on whether one or both persons, A and B, are in infancy. To make an arbitrary choice, I will limit my discussion to the interpretation in which A is an adult and B is a small child. The sentence "Mommy will come there tomorrow morning" is of sufficient structural complexity and maturity that somebody young enough to use the word *Mommy* in talking about his own mother would probably not be expected to form this sentence. (Here, of course, I am using information that is not completely linguistic.)

Mommy is not just a word used in a text that enables us to know something about the speaker and hearer. It is also the subject of the verb *come* and hence is the person who is going to do the traveling. The words tells us that she must be related to somebody by the mother/child relationship. The speaker did not say "your mommy" or "my mommy," but just "Mommy." A child would say "my mommy" when talking to somebody who was not in his own family, but just "Mommy" when talking to somebody who lives in his own household. We are assuming here, however, that A is an adult and that B is a child. Since we know that an adult talking to a child will tend to identify people from the child's point of view, we assume that the person intended is not the speaker's mother, but the child's mother. Since the expressions is "Mommy" and not "your mommy," we assume that the speaker and the child are members of the same household. We must keep A's identity indefinite, since we do not know from this evidence alone who in B's household A might be. A might be B's father; A might be B's mother, referring to *herself* as "Mommy"; or, of course, A could be an elder sibling, an aunt, a grandparent, or a household servant.

The result of this kind of figuring is that we can add to the description of the scene, in the version that we are working with, some 'footnote' information about the relationships among our cast of characters. Some of this is the information that D is B's mother, that A and B are members of the same household, and that B is a small child.

Now let us return to the word *come.* It presupposes movement to a destination that is in some way connected with the conversation participants. The conditions are simple, but they are hard to formulate. The possibilities include the situation where A, the speaker of the sentence, is at the destination F; this, for example, is the interpretation we would clearly get for a sentence like "Come here, please." A second possibility is that F, the destination of the movement, could be the current location of B, the addressee; this is the interpretation we would necessarily get in a sentence like "Wait for me, I'll come there right away." A third possibility is that F is the location of the speaker at the time of the mover's arrival; this is the interpretation we provide for a sentence like "I'll get there around two, and he'll come there around two thirty." The fourth possibility that I have in mind is that F could be the location of the addressee at the time of the mover's arrival; this is the interpretation forced by "As soon as you get there, he'll come and help you."

The verb *come*, in other words, allows all of these possibilities, plus a few others, all related in some way or another to the location of one of the participants in the speech act. So far, then, the description of the scene schema for our sentence will have to leave all of these as possibilities.

But now notice that the next word is *there*. For a place to be referred to as *there*, it must be a place where the speaker is not located at the time he uses the word. Therefore, one of the four possibilities that we had in mind earlier gets ruled out: the possibility that the speaker, A, is located at the destination, F, at the time of speaking the sentence.

The various possibilities depend, of course, on who A, B, and D are–in particular, on whether A happens to be D, that is, on whether the speaker is referring to herself when she uses the name "Mommy." I suggest that the scene we create for the sentence be the one that is limited to that possibility. One possible contextualization of that situation has the mother talking to the child over the telephone.

Now the word *tomorrow* locates the moving event with respect to the speaking event, the moving event being one day future to the speaking event. Furthermore, the moving event is located with respect to the "morning" subperiod of that next day. The method of communication used by A in talking to B has to be one in which they can share the kind of temporal reference point necessary for the use of deictic time words like tomorrow; as suggested before, the possibility of communication by telephone fits this condition.

Given these assumptions, then, we now know where to locate, in time, the *move* scene: in the *morning* part of *tomorrow*. But to be sure about this guess, we ought to try to unpack the notion of motion. A journey is described by specifying the starting point, the itinerary, and the destination. We need to be clear about whether the morning period includes the whole journey or only part of it; that is, we need to know whether to paste into the morning space in the strip the whole moving scene, or just the beginning or the finishing end of it.

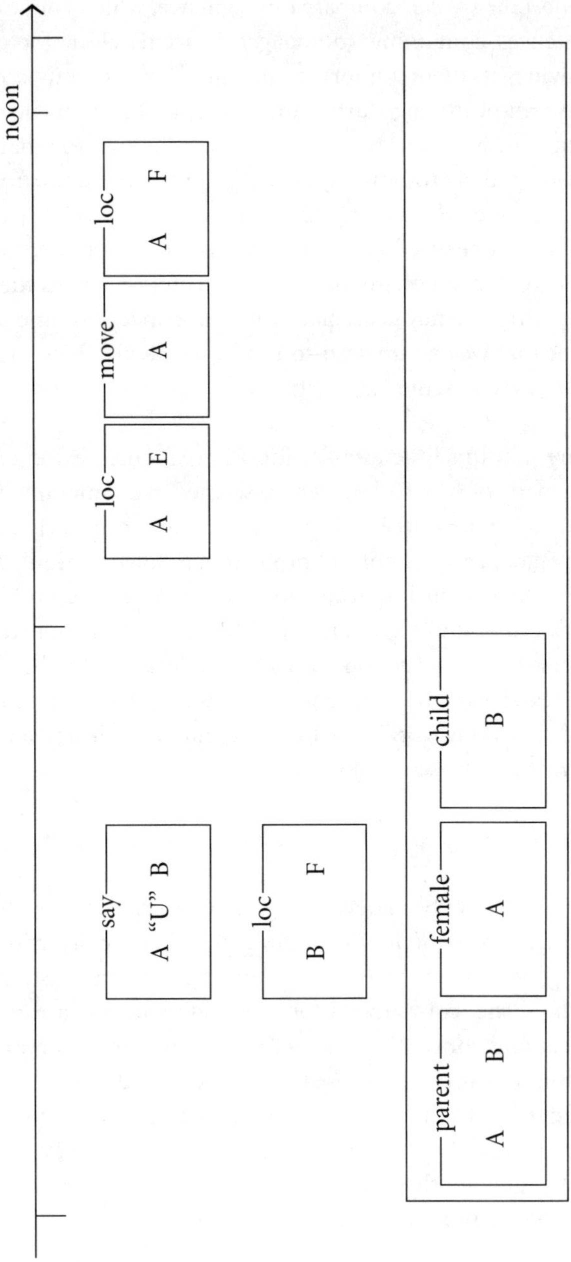
noon
loc
A F
move
A
loc
A E
say
A "U" B
loc
B F
parent
A B
female
A
child
B

Fig. 17

To see if we can determine how much of D's journey is to be located within *tomorrow morning* we can compare our sentence, which contains *come*, with similar sentences containing *go* and *get*. It seems clear, for example, that "Mommy will get there tomorrow morning" gives us a scene with the mother's *arrival* occurring during the morning, but with no commitment about the time of her departure; and it seems equally clear that the sentence "Mommy will go there tomorrow morning" locates her *departure* time within the morning period, but is compatible with a later arrival. My impression is, though I have not checked this out with more than a few speakers, that the sentence "She will come there tomorrow morning" requires the understanding that the entire journey takes place within the specified time period.

The scene that we end up with–one of the several scenes compatible with this sentence–is represented as Fig. 17.

11. I am sure that this little exercise must have seemed extremely trivial and pointless to many of you, and must have seemed like something that concerns itself with details that a decent theory of semantics can easily postpone until certain more global and important problems are solved. I think, however, that the problems that seem important to an analyst are the problems that are hard to work out within a given model. This sentence demonstrates the ways in which the scene-construction model forces the analyst to be clear about, or at least to seek clarity about, a range of questions that can easily escape the attention of somebody working with a different notation and a different set of assumptions about the scope of semantics.

LECTURE FOUR

0. The notation I have been developing in these lectures is one that is capable of representing succession in time, simultaneity in time, selection from a contrast set, 'stage settings' or aspects of a scene that remain constant through the period of the scene, conventions for cross reference in cases of agreements and contracts, and conventions for embedding in a scene a representation of the very communication act whose text the scene analyzes.

In this final, brief lecture I plan to do three things: introduce the problem of representing models of a world that are in conflict, survey some of the terminological problems these proposals give rise to, and review, in terms of these proposals, a number of traditional concepts and issues in semantic theory.

1. One of our representational problems is that of showing how one person's model of the world can include information about another person's model of the world. The representational problem is one of scene embedding. The comic-strip tradition provides conventions that might be put to use, like, for example, the balloon that represents a character's speech or thoughts, but the problem of embedding becomes serious precisely because there can be embeddings of unlimited depth. To analyze a sentence of the form "Julie thinks that her dad was honest," we will need to express the notion that the speaker's model of Julie has it that Julie's model of her father has it that her father's model of the world is compatible with the models of the world that match his speech utterances. The comprehensibility problem of a balloon over a character in a balloon over a character in a frame is probably obvious without illustration. I will say no more about that problem here.

2. The other set of cases with which I shall interest myself has to do with the vocabulary of human emotions. Some emotion words, I believe, can be captured with single scenes, in the sense that the emotional states they name are relatively simple and have conditions that can be contained in a moment in time. Some of the single-scene emotions are those named by the words *excitement*, *anger*, *peacefulness*, and *joy*.

Multiple-scene emotions are those that entail a history; for some emotion-naming words, the word is appropriately used only if it can be understood that the emotion-experiencing event is a stage in a history. One linguist who has emphasized this aspect of emotion vocabulary is Anna Wierzbicka (1972, pp. 55–70) from whose work some of these examples are adapted. Wierzbicka's paradigm emotion description takes the form "The way someone feels when ...". In our terms, the emotion word names the final scene in a history, with the assumption that the experiencer of the emotion is reacting to this history in a normal or prototypic way.

A prototype characterization of *disappointment* might go something like this: *Disappointment* describes the way a person feels who has wanted something to happen, who has had reason to believe that it would happen, and who has discovered that it did not happen the way he expected. In our strip notation, the first frame has A wanting B to happen and expecting B to happen by the time represented in the next frame; that next frame shows A realizing that B is not going to happen; and the third frame has A reacting to this experience. The arrow links the word *disappointment* to this last frame.

Here are some other words that can be defined along these same lines. We can describe *frustration* as the way someone feels who has wanted to do something, who has tried repeatedly to do it, and who has just realized that it cannot be done. *Surprise* can be described as the way someone feels who has believed A and who has discovered non-A. *Suspense* can be defined as the way somebody feels who wants to know whether some state of affairs will come about, who realizes that he must wait longer before he can know whether or not it will come about.

In all of these cases, we have emotion words whose definitions require an understanding of conflict between the worlcl and one's model of the world. Either the world that one desires is not the same as the world that one sees, or the world that one has believed in turns out not to be the world that one is experiencing.

3. It is clear from these examples that my previous accounts of scene semantics have not provided all the distinctions that need to be made. In particular, I have given the word *scene* too much work to do. Here are some of the distinctions that I think are going to prove to be important.

In the first place, we need to recognize the real-world scenes in terms of which people have learned categories and distinctions, and in terms of which people have acquired their original awareness of the objects and experiences that the world has to offer, as well as the real-world scenes that are the contexts and causes of ongoing perceptions and behavior.

Secondly, there are memories and distillations of real-world scenes in people's minds, possibly restructured in ways provided by their participation in a particular community, possibly with some aspects of them forgotten or suppressed and others enhanced.

Thirdly, there are schemata of concepts, stereotypes of familiar objects and acts, and standard scenarios for familiar actions and events that can be spoken of independently of given individuals' memories of experiences.

Fourthly, there is the imagined scene of the speaker as he is formulating his text; and fifthly there is the imagined scene of the interpreter as he is trying to construct a model of the world that matches the text he is interpreting.

And lastly there are the sets of linguistic choices that a given language provides and the ways in which these activate or are activated by particular conceptual schemata.

These are fairly confusing notions, and I among many have never had a time when I felt I had a good conceptual grasp of the whole collection of concepts. The area of confusion is the area for which workers in language understanding projects within artificial intelligence have developed terms like *scene, frame, schema, description, template, scenario, prototype, module,* and *model.*

One possible set of terminological choices is the following. We can use *scene* to refer to real-world experiences, actions, objects, perceptions, and per sonal memories of these. We can use *schema* to refer to one of the conceptual schemata or frameworks that are linked together in the categorization of actions, institutions, and objects–such as the collection of notions we had to recognize for characterizing the commercial event–as well as any of the various repertories of categories found in contrast sets, prototype objects, and so on. We can use *frame* to refer to the specific lexicogrammatical provisions in a given language for naming and describing the categories and relations found in schemata. And we can use *model* to refer to either somebody's view of the world or the view of the world that an interpreter builds up in the process of interpreting a text. A *text model* can be thought of as the assembly of schemata created by the interpreter, justified by the interpreter's knowledge of the frames in the text, which models some set of possible complex scenes.

The integration of these concepts can be talked about in this way: from experiences with real-world scenes, people acquire conceptual schemata; in the acquisition of schemata, sometimes items from language frames are learned for labeling these and their parts; *words* from a language frame activate in the mind of the user the whole frame and the associated schema; the schemata can be used as tools or building blocks for assembling, on the basis of the words in a text, a text model–i.e., a model of the world that is compatible with the text.

We can think of a schema, as here defined, as a standard set of conditions, or as a conceptual framework, that characterizes ideal or prototypic instances of some category. As human beings, we can interpret an experience if we can succeed in assigning some sort of conceptual schema to it, that is, if we can locate the experience as an instance of the schema. In many cases this assignment is done "to a degree"–that is, the scene at hand may not match the schema exactly, but it approximates it to a degree.

We can talk about an experience if we can assign a schema to it and if we know the linguistic frame that fits that schema. People can understand our talk if their linguistic repertories activate the same or comparable schemata and if their experiences in acquiring these schemata are comparable to ours. Our interlocutors can understand what we are saying if they can assemble the schemata we have introduced into a model of a possible scene that matches the model we wished to impart to them. The process of interpreting a text, in short, can be thought of as involving a set of procedures for constructing a coherent model of a possible world.

One of the first types of objections to a world-building or image-building way of talking about text interpretation has to do with the treatment of negative sentences. How, one might ask, are we to imagine a world or scene that matches the sentence, "There are no elephants in my kitchen today." It is easy enough to imagine an elephant-free kitchen, but there is not likely to be anything about that kitchen that makes it fit the given sentence better than it will fit any of the infinitely large set of other negative sentences compatible with the same scene. My answer, of course, requires an understanding of the difference between the internal and the external contextualization of a text. The scene, in other words, has to contain an utterance of the sentence, and the understanding we seek is an understanding of the kinds of situation in which it would be appropriate to produce that particular negative sentence. These situations will involve, in the trivial case, the occurrence of the sentence as an example in a philosophy class or a linguistic class during a discussion of imagery theories of meaning, or, in the more interesting case, a situation in which somebody has already introduced into the conversation the possibility of there having been, or of there being in the future, some number of elephants in the kitchen being referred to. For some negative sentences, in other words, the scene that the interpreter constructs contains not only properties appropriate to the sentence itself but also properties connected with the context in which the sentence could be meaningfully produced.

4. My final task is to show that the way of talking about meaning that I have introduced here is capable of offering something illuminating, or at least something not clearly less illuminating than more traditional semantic theories have allowed, about a number of traditional semantic notions.

4.1. Let me begin with *ambiguity.* We can say that a sentence is ambiguous if some linguistic form in the sentence is assigned to two different frames. The sentence "My desk drawer is ten inches deep" is ambiguous because the word *deep* is found in two different contrast sets compatible with the measurement use we find here. In one of its senses it is contrasted to *shallow*, the two making up a pair of polar opposites, with *deep* being the word used in giving measurements. In this sense the sentence can be understood as saying something about the top-to-bottom dimensions of the interior of the drawer. In the other of its senses it is contrasted to *wide*, the two making up a pair of words for measuring the front-to-back as opposed to the side-to-side measurements of any object having a roughly oblong shape. In this sense the sentence can be understood as saying something about the front-to-back dimensions of the drawer.

4.2. A sentence is *vague* in the extent to which the transition from the scenes that clearly match and those that clearly do not match the frame it activates is gradual. The sentence "My desk drawer is deep" is both vague and ambiguous: vague in that there is no clear boundary between being deep and being not deep, in either of the two senses of *deep.*

4.3. A word is relatively *general*, rather than *specific*, if the description of the associated scene is relatively less detailed than whatever word it is being compared with. It follows from this description that the number of detailed scenes that match it is greater for a general term than for a more specific term; yet that condition alone does not capture the idea of generality. The word *calf* can be used to designate the young of a cow, a hippopotamus, or an elephant; but I would not be inclined to say that it is thereby more *general* than it would be if it could refer only to the young of a cow. A single frame characterization of *calf*, if that were possible, would have to include the information that the word is not applicable to the young of horses, wolves, etc.

4.4. Two words are *synonyms* if different lexical choices are available for the same element in the same frame. Perhaps in this sense *eye doctor* and *oculist* are synonyms; and quite likely *furze* and *gorse* are, too.

More interesting than *complete synonymy* is *partial synonymy.* Words are partial synonyms of each other if parts of the frames in which they occur are identical, other parts being different. *Promise* and *guarantee* are partial synonyms. Both agree in committing the agent to the future occurrence of whatever action or event is the content of the promise or guarantee; but *guarantee* adds to this the understanding that the receiver of the guarantee has rights against the guarantor should the promise fail. *Signature* and *autograph* are partial synonyms, the latter differing from the former in the details of the scene having to do with the motives or interests of the person for whom the signature was obtained.

4.5. The concept of *selection restrictions* or *co-occurrence constraints* can be spoken of in terms of the linguistic frames that are associated with given schemata. Some words are limited to the kinds of schemata they activate and are constrained, therefore, to appear with the other words that belong to the frames matching such schemata. The notion of selection restriction can thus be formulated in terms of the properties of schemata and frames, and not in the usual way, i.e., in terms of the pairing of inherent features and distributional features on lexical items.

4.6. The concept of *antonymy* is more or less captured by the concept of contrast set discussed earlier in connection with what I was calling *selection.* If a frame allows one choice from a set of mutually exclusive categories, these can be thought of as forming an antonymy set. The usual sense of *antonymy* is captured by the situation in which an antonymy set contains exactly two member categories.

4.7. The concept of *category boundaries* for words can be reexamined in terms of the concept of the *prototype.* The research of Labov on the boundaries of notions like *cup, bowl, glass,* etc. (Labov 1973), can be illuminatingly discussed in terms of the strategies by which subjects make use of prototype elements in schemata. The schema for the set of things one can expect to find in one's kitchen includes a prototype cup, glass, and bowl. When subjects make judgments about some sort of monster cup presented to them in an experimental setting, they are not really deciding on the boundaries of the category of cup, but rather they are trying to decide whether the new object is close enough to being a kitchen vessel to deserve comparison with the kitchen-cabinet schema in the first place, and then to decide if it is close enough to one or another of the prototypes to be classed as an instance of that type rather than of some other. Experiments of category boundaries are really experiments that test people's strategies for extending from a prototype to a nonprototypic instance of a category.

4.8. The concept of *semantic field* can be captured by appealing to the notion of schema, and the allied concept of *vocabulary field* can be identified with the notion of frame and with various linkages among frames. The human color schema identifies the semantic field of color terms; the commercial event schema underlies the vocabulary field of buying and selling. And so on.

The schema and the associated linguistic frames for the human body provide for the field of body part names and the vocabulary of body positions and body movements. I believe that the kinds of definitions that are sometimes given for certain kinds of body activities can only be made comprehensible if we assume the existence of a body schema and if the body schema is interpreted, as I have been suggesting here in terms of *prototypes.* Eugene Nida (1975) has offered analyses of certain body activities that can be adapted as follows: Let L stand for one human foot and R for the other, and let an occurrence of these symbols mean that the designated foot is on the ground; then let O stand for a period when neither foot is on the ground. Then we can define *hop* as LOLOLO..., *skip* as LOLOROROLOLORORO, and we can define *run* as LOROLORO..., and *walk* as LRLRLR... . Now I would suggest that it is only if we accept a prototype notion of human locomotion that these definitions are acceptable. I am sure, in other words, that it is possible to meet the conditions of the definition of any one of these and yet be doing something that would not qualify as an instance of the type of locomoting named by the word being glossed. I think, in other words, that only if we already have an idea of how people typically move about, and only when we keep in mind that set of options, can these definitions correctly separate the various types from one another.

4.9. The semantic notion of *converse* can be captured as follows: If there are two predicating words that belong to the same frame but occur in syntactic structures that require that the entities of the associated schema are mentioned in different orders, then those two words are converses of each other. *Buy* and *sell* are converses in this sense, as are *taller than* and *shorter than*, or *husband* and *wife.*

4.10. I believe that the traditional linguistic notion of the *core meaning.* (Grundbedeutung) of a linguistic form can be usefully discussed in the terms I have been setting up here. The goal for the linguist who tries to construct or discover the core meaning of a form is to find one formulation of its uses that will cover all cases. The goal is to reduce to a minimum all appearance of polysemy. My feeling is that in general this goal is not a good one for linguistic semantics. To return to the example of *calf,* I would rather say that it is a member of several linguistic frames, and perhaps add that in each of these frames its relation to the associated schema is similar, than to try to formulate as the meaning of *calf* some single statement that will exactly cover all cases of its use. I think, in other words, that the most useful information about a lexical item is the set of frames in which it plays a role and the position that it occupies in each of these frames. Whatever subregularities are lost by making this choice can be presented in the form of an historical account of the evolution of the language system, that is, in terms of the extensions of words from one frame to another in the history of the language.

Suppose, for example, that somebody decided to identify the core meaning of the word *short* independently of whether it was used as the opposite of *long* or as the opposite of *tall.* Since a semantic account of English would still need to provide the information that the word appears in just these two contrast sets, nothing could be added to our understanding of the language by this decision. Consider, as an exercise, the series of adjectives *high-low-tall-*

short-long-wide-deep-shallow. It happens that each of the surrounded words in this list belongs to a different contrast set with the word to its left and with the word to its right. For example, we can speak of low and high clouds, tall and low buildings, short and tall people, long and short wires, wide and long tables, deep and wide plots of land, shallow and deep swimming pools. The decision to represent each of the internal adjectives in the list with a single statement of its meaning would completely obliterate the associations with the large number of distinct contrast sets just exemplified, and would contribute very little to our knowledge of how the English vocabulary is organized.

4.11. A frequent topic of discussion among semanticists is the issue of where and how to draw the line between linguistic information about the meanings of words and real-world information about the properties of things. This issue usually takes the form of the question "What is the difference between a dictionary and an encyclopedia?" The famous Spanish Academy definition of *dog*, as the species in which the male urinates by raising one leg, or the common dictionary definitions of *left* or *right*, which speak of the side of a person facing south or north when that person is facing west, clearly are not conceptual analyses of their definienda, but rather serve as recognition tests for people who need to make sure what kinds of things the words designate. It is frequently assumed by linguistic semanticists that the linguist's job is to determine the purely linguistic information about word meanings and that a distinction between a dictionary and an encyclopedia can in principle be established. A more realistic view might be something like this: there are things in the world, there are typical event types that one can observe in the world, and there are institutions and cultural values that make human endeavors interpretable; for a very large part of the vocabulary of our languages, the only form a definition can take is that of pointing to these things and actions and institutions and indicating the words used for naming and describing parts and aspects of them.

I do not assume that the role of the lexicographer and that of the encyclopedist cannot be distinguished; but I also do not think that everything that a native speaker knows about the meaning and use of a word can be encapsulated in a dictionary entry. The most useful practical dictionary will in many cases simply appeal to the reader's knowledge of the world, giving enough information to enable the user to acquire further information elsewhere if he fails to understand sentences containing the word and the dictionary definition fails him.

It seems clear, at any rate, that any attempt to relate a person's knowledge of word meanings to a person's abilities to interpret texts will have to recognize the importance of nonlinguistic information in the interpretation process. We get clearly different interpretations from the sentences "The fly was on the wall" and "The cat was on the wall," just because we know different possibilities for stable positions for these two kinds of animals and because we know that the same word–*wall*–can be used to refer to a vertical surface of a room or building and to a high-relief boundary around a place. Generally this kind of disambiguation is thought of as a use of semantic competence; but in this case it surely involves information of the kind that cannot be sensibly incorporated in the definitions of the associated words.

4.12. Another semantic notion for which the difference between linguistic and encyclopedic information is relevant is the concept of *metaphor.* The act of interpreting a metaphor requires an understanding of the kind of social-interaction schema within which the speaker can expect the hearer to do some special construing of what the speaker has said, and it requires an understanding of how the detection of a mismatch between the literal meaning of what has been said and the text world that is 'under construction' can trigger these special construal efforts. If we hear something like "Harry is a pimple on the face of the community," we do not use peculiarly linguistic information to interpret what has been said. We know enough about people, pimples, and communities to know that a coherent scene cannot be constructed out of that sentence taken literally. From that detection of a mismatch, we know that we should make use of the psychocultural information that people are embarrassed by and want to get rid of a pimple, and we assume that the speaker intended us to believe that the members of the community have feelings like that about Harry.

A common approach in linguistics to an explanation of the metaphoring process has been to determine the criterial features of a word and to devise a formalism for showing how the features appropriate for one word can be superimposed on the collection of features associated with a second word or group of words in the same construction. I believe that most cases of metaphor, including the most interesting ones, are not like that at all.

4.13. The concept of *abstraction* has always been a difficult one to point out and to be clear about in linguistic semantics. Abstractions are not names of things, but they are not simple predicates or predications either. They are somehow to be taken as the names of larger situations, and they frequently occur in sentences in which some comment is being offered on that larger situation. In the kind of notation I have been proposing here, the concept of abstraction will be represented, not by an arrow to any particular object in a scene, but by an arrow that points to the scene as a whole. Furthermore, in the extent to which a concept is an abstraction, the schema that expresses it will not be fitted into the time line that goes with the text model, but must be taken on the side as a kind of variable.

I have no new proposals on the nature of abstraction, but I can at least point out that it may be necessary for semantic theory to recognize many different uses for, or points of view in the use of, a single abstract concept. For example, in the prototype abstraction *charity*, we know independently of any specific use of the word that somebody gave something to somebody, the giver was not under any obligation to do what he did, the receiver as a result of the charitable act undertook no obligation to the giver, and the giver believed that what he did was of benefit to the receiver. Once we have characterized a situation like that, we sometimes need to call on fairly complicated procedures for interpreting sentences containing the word, procedures that do not merely point to the situation as a whole.

For example, when we interpret a sentence like "Charity is a virtue," what we have in mind is that the giver in an act of charity is judged by the speaker to be good. When we interpret a sentence like "Charity is degrading," we figure out that the receiver of the gift is being described as feeling insufficient for needing and undignified for accepting that kind of help. When we come across a sentence like "Charity is unnecessary in an ideal society," we are saying of this 'ideal society' that there are no situations in it that fit the charity schema. And a sentence like "He did it out of charity" reveals something about the giver's inner life at the time of the gift–that is, we are being told that the giver felt no externally imposed obligation to do what he did, that he expected no reward, and that he felt that what he was doing would help someone.

4.14. One important question that comes up frequently in lexical semantics is that of the motivation for *lexicalization*. What does it take, the question is, for something to be expressed as a single word rather than more analytically? In specific terms, when would somebody say *kill* rather than *cause to die*?

In our understanding of language in general, there seems to be a schema for lexicalization the sense of which is that the act of lexicalizing something is the act of presenting it as an established category of human thought. If a lexical item exists, in other words, it must exist as some part of a frame and must correspond to some part of a schema. The English word *vegetarian* does not merely designate a person who lives exclusively on vegetable food. If everybody did, or if all creatures on earth did, there would be no particular need for such a concept, and hence for such a word. There is such a word because there are many people who eat meat, and by contrast there are people who deliberately do not eat meat. It is by virtue of this contrast that a word like *vegetarian* can have a function in our language.

Since everybody eats, there is no word **eater*; but there is a place in our language for a word like *chicken eater*, precisely because we can easily imagine a difference between people who eat chicken and people who do not. My remark that "there is no word **eater*" can easily be challenged. But the point is that for there to be an occasion for such a word, there would have to be, or to be imagined, some situation in which being an eater was importantly distinguished from not being an eater. Presumably in a community facing starvation there can come to be certain principles for deciding who will eat and who will starve; and in such a situation a contrast between eaters and non-eaters becomes a meaningful one.

4.15. Instances of *functional shift* can be seen as instances of a special kind of lexicalization process. There are various conditions under which, in English, a noun can acquire a verbal use. One such word is *hammer*; another is *spoon*. A simple rule for describing this derivational process might be that a noun that names the instrument with which some action can be accomplished can come to be used as a verb designating the performance of that action with that instrument. It is important to realize that the verb *hammer* does not simply refer to hitting something with a hammer, but rather to hitting something, possibly with an implement that is not actually a hammer, in the manner in which and for the purpose for which hammers are typically used. Similarly, to *spoon* something is not simply to transfer it from one place to another by means of a spoon, but to do so by using a spoon in the most typical way in which spoons are used, in particular, by using a spoon in the kinds of actions for which spoons were invented.

5. I have tried in this last lecture to say something about what it would take to push the 'scenes' model further than we were able to in the first three lectures; I have tried to lay out a number of distinctions that a semantic theory *cum* text interpretation theory of this kind needs to take into account; and I have tried to show that this approach to semantics when compared with the more traditional formal models of semantics does not leave us obviously less able to talk intelligibly about traditional concepts and problems in semantic theory.

I would like at this time to emphasize the tentative nature of whatever conclusions I have reached in these lectures, as well as the tentative nature of my commitment to the terminology and notation that I have introduced here. In a great many conversations I have had with other scholars who use terms like *frame, schema, scene,* and *prototype*–cognitive psychologists, philosophers, linguists, and computer scientists–I have been impressed first with the variety of conceptualizations that underlie these terms, and second with the reality of subtle changes in my own use of these terms as a result of these conversations. These are not good signs. The possibility of my being misunderstood equals in seriousness the possibility of my not remembering one month what I meant one month before. Those who have come with me this far need to be given these cautions.

7

On the Organization of Semantic Information in the Lexicon

This chapter appeared as an invited contribution to a "Parasession on the Lexicon" in the Chicago Linguistics Society meeting of 1978. It is a kind of hit-and-run list of issues of meaning and interpretation related to the lexicon; it makes me think today of what might happen if a stand-up comedian is given one chance to perform before a particular house and tries to include all of the jokes he knows.

Whereas the preceding chapter gives a broader view of sentence and text meaning within which to see the role of case structures, this chapter makes it clear that case descriptions limited to argument-structure bearing words is a relatively small fraction of the full range of semantic information found in a lexicon.

The basic message of this chapter is that the complete lexicon of a language incorporates a vast variety of lexical types calling for an equally large system of descriptive frameworks. These must allow for recognizing the relation of a word's meaning to its morphological structure, the importance of lexical items that are larger than single words, the difference between naturally emerging meanings and meanings stipulated for special purposes (terminologies), the difference between uniquely coined and systematic meaning extensions, as well as the number of ways in which lexical systems can differ across languages.

1.

My intention in this paper is to examine the various ways in which one might speak of the structuring of semantic information in the lexicon of a language. I begin by raising for consideration the following questions:

(1) How is lexical information in the formal linguistic description to be organized and presented? What formal properties should the entries in a lexicon have, and how does semantic information make up a part of this?

(2) How does a lexicon, taken as a closed set of lexical items, cohere? That is, what is changed if one lexical item is removed from the list or if a new one

Originally published in 1978

is added? Are there vocabulary domains where this makes a difference and others where it does not?

(3) How are lexical items interrelated? In particular, to what extent can the description of one lexical item make reference to specific lexical items? When does such cross-referencing appear to be necessary, and when is it merely convenient?

(4) What sorts of stratification of the vocabulary of a language need to be recognized? Should different strata be marked as appropriate to different contexts, styles, registers, or discourse topics? How do languages differ with respect to the extent, character, and function of such stratification?

(5) How do we assure ourselves that the organization we see in the description of a lexicon reflects realities in the language rather than the consequence of assuming some particular methodological stance?

(6) What sorts of notions or semantic configurations get lexicalized, that is, get realized as individual lexical items? Are there universal constraints on lexicalization patterns? Are there characteristic lexicalization patterns; distinguishing different languages from each other in systematic ways? How can these be most revealingly described?

(7) How are semantic properties parcelled out among different classes of words? How do these assignments differ from language to language?

(8) How in general do languages differ from each other with respect to the ways in which their lexica are structured?

In this paper I will consider the question of what constitutes lexical semantic information, and I will discuss the connection between the analysts' theoretical or methodological commitments and the kinds of structures or appearances of structure that emerge from their studies; I will then review some of the ways in which languages differently organize semantic information. After that I will discuss vocabulary stratification, degrees of semantic transparency in word forms, differences in the ways of assigning semantic information to morphological material, and the structural properties of different types of lexical fields; and I will make the suggestion that different domains of vocabulary require importantly different sorts of semantic description. Lastly I will examine briefly some characteristics of the language of law, both as an exemplification of some of the principles described earlier and as a backdrop against which to view the semantic structures of ordinary language.

2.

A decision to study lexical semantic structures presupposes the answerability of the questions what is lexical and what is semantic. Both of these are subtle questions. Scholars disagree on how much the set of possible interpretations of a sentence can be attributed to the meanings of constituent words, the imposition of meanings by grammatical categories, the operation of semantic construal principles, such as those that would account for the interpretation of non-conventionalized metaphors, the operation of principies of conversational cooperation, or, indeed, knowledge about the non-linguistic aspects of the world. In general these problems have been slow in yielding to the analyst's probe.

On the question of what is properly 'lexical' there are problems of the appropriate size level for the entries described in the lexicon; and this, I believe, cannot be resolved merely by accepting or constructing a definition of the word. We need to distinguish the idiosyncratic from the general in separating linguistic knowledge into the lexical and the grammatical, and that effort will result in our recognizing as 'lexical' both things that are larger and things that are smaller than ordinary words.

In particular, we have to face the inconvenient reality that a number of expressions in any language have to be viewed both as lexical items and as entities having grammatical structure on a level higher than that of word-formation. I have in mind, not only the much-discussed 'idioms,' which probably make up only a small small portion of the total 'phrasicon,' but also the vast repertory of fixed phrases, clichés, speech formulas–in general, all conventionalized ways of saying things–that a speaker acquires independently of the process of learning the grammatical rules of the language.

Thirdly, there is indeed the problem of distinguishing the lexical from the grammatical. There are semantic observations which have competing lexical and syntactic accountings. Such disagreements can be seen as due to different assumptions about legitimate levels of syntactic representations of a sentence. Here is an example of one such problem.

About a year ago the *San Francisco Chronicle* had a story about some patients in a psychiatric hospital who assisted one of their fellow patients in the birth of her child. (The doctors hadn't believed her claim that she was pregnant.) The headline of he story was

Mental Midwives

While granting that journalese uses very special codes whose nature may have nothing to do with what real languages are like, we still might find it useful to talk about what the author of this headline did. Shall we say that he or she intended the word "mental" in one of its accepted senses and produced the phrase "mental midwives" as a kind of newspaper headline abbreviation of a larger syntactic structure, or shall we say that the journalist created a new sense for this word, a sense paraphrasable as 'being a mental patient' or the like. Once we make up our minds about that, what shall we then say about the word "mental" in the phrase "mental patient" itself? Shall we say this time, on the grounds that it is an accepted and familiar use, that the word "mental" has as one of its senses the meaning 'suffering from a mental disease , or shall we say here too that the phrase "mental patient" is a syntactic abbreviation of something else? Exactly analogous question can be asked about a very large set of locution types in English. School districts often have "retarded programs"; cities sometimes have "topless districts"; etc., and a semanticist needs to be able to say something about the function and meanings of the adjectives in these expressions. (Cf. Leech 1974, p. 223)

A fourth confusing element in doing lexical semantics comes from the fact that the purely descriptivist approach sometimes gives us nowhere to go. We frequently need to set aside the commitment to construct semantic descriptions that account for attested usages in a straight way. In many situations, I think, we need to regard our observations as comprising mistakes, performance errors, data that need to be explained, or course, but not in terms of semantic description alone. In the *Encyclopedia Britannica* article on "Comparative Jurisprudence" which I once found myself reading, I came across the following context for the word "conception":

> Many of the simpler peoples see no causal connection between conception and childbirth.

It doesn't make sense to ascribe to a people the concept 'conception' without at the same time attributing to them an understanding of this event as a stage in the creation of a new living being. I suspect that the author must have chosen the word "conception" as a euphemism for sexual intercourse. I think, in other words, that even for this author it would be wrong to regard the usage as involvlng a proper sense of the word "conception." Unfortunatey, the decision to dismiss a putative semantic observation as a performance error is a decision we make at some risk, since many meaning changes originate in precisely this way.[1]

3.

An important question to ask in any area of linguistics is that of the difference between structure discovered and structure imposed. We need to keep in mind both the danger of choosing a model or method which will prevent us from detecting structures which the language is willing to show us, and that of choosing one which creates structure where there is none there from the start.

A key concept in the theory of lexical semantics proposed in Bendix (1966) is what he calls the "minimal definition" principle. This principle requires that the semantic description of each vocabulary item in a language should consist of exactly those features which will distinguish it from every other vocabulary item in the language, and no more. The result of constructing a set of lexical descriptions along such lines is going to be the kind of perfectly cohering minimal network that is most pleasing to those who work within the structuralist tradition. Any new word added to the language could require revisions in the semantic specification of possibly great numbers of other words in the language; and any existing word removed from the list could require massive reshapings of the descriptions of other words. A commitment to this principle for its own sake makes it impossible to distinguish those lexical domains in which this sort of structuring principle is valid from those where it is not.

A critical defect in this approach is that, if in the normal way of talking some word has a meaning which requires or invites a particularly rich and detailed accounting, much of that information would have to be left out of the lexicon in principle, just in case it is redundant from the point of view of the minimal definition principle.

Some scholars wish to recognize redundancy-minimizing as a proper goal for semantics and at the same time to justify inclusion of all of the relevant semantic information for each word. Bierwisch and Kiefer (1969) find themselves distinguishing the central meaning of a word from what they call the 'periphery', the latter containing those parts of the meaning of a word which do not figure in the 'structure' proper. The point of my mentioning it is that

1. I have made a similar point regarding journalists' and police officers' use of the words "allegedly" and "suspect" in Fillmore (1972).

this distinction is a natural one only if the redundancy-minimizing goal is taken as a proper one for semantics.

Related to the minimal definition principle is the position that for each word a single core meaning should be found that covers, if at all possible, all of the word's significations. This is the polysemy-minimizing position that is associated with such scholars as Roman Jakobson and William Diver, though in their work it has figured mainly in the descriptions of the meanings of grammatical morphemes such as those signalling cases and tenses. I don't have much information to guide me on how a commitment to such a position would lead one to proceed in normal lexical description except for Jakobson's reported proposal for simplifying the multiple-meaning account of the word *bachelor*, first made famous in linguistic circles by Katz and Fodor (1963). The word *bachelor*, we have learned, can mean either an unmarried man, or a young male fur seal without a mate during the mating season, or a knight serving under the banner of another knight. Jakobson's single formulation, that is said to make it unnecessary to regard the word as polysemous, is "unfulfilled in a typical male role".

If I understand this position correctly, it would be important to those who hold it to regard *short* as having a single description that covers both its opposition to *long* and its opposition to *tall*, and a single description that covers *wide* both in its relation to *long* and in its relation to *deep*.

It is my opinion that a commitment to this position would obscure certain kinds of lexical structure where they exist–in particular that of contrast pairs that by coincidence share a single lexical form.[2]

Another methodological commitment that could lead the semanticist astray is the search for–or rather, the belief that one has found–the ultimate restricted paraphrase language for semantic description. While I think the exercise of constructing paraphrases with these systems is a useful one, forcing the analyst to recognize semantic properties that could otherwise go undetected, I suspect that it may be wrong to assume that precise conclusions about the semantic structuring of lexical items can be drawn from products of such exercises. Words, in such accounts, can be shown to differ from each other with respect to the relative complexity if their description in the associated paraphrase language; but I think it is not always necessary to believe that the real relative complexity of the words has therefore been revealed.

2. By *coincidence* here I mean from the point of view of the functioning of the contrast pairs in the current language. There may well be explanations in the history of the language that would make the double function of, say, the English word *short* seem not at all *coincidental*.

I have urgued elsewhere (Fillmore 1975b, p. 141) that such notions as the relative complexity of related vocabulary items can be given different appearances depending on the choice of primitives. If the central primitive terms in a kinship semantics are *parent*, *male* and *female*, *father* and *mother* are complex terms, both equally complex, while *male*, *female* and *parent* are all simple terms. If, on the other hand, the primitives chosen for the system were phenomenologically more primary ones like *childbearing*, *impregnating*, *inheriting a body equipped for one of these*, then *mother* would be the simplest term in the set, *male*, *female* and *father* would be more complex, and *parent* would be the most complex of all, comprising a disjunction of the descriptions of *mother* and *father*.

The same point can be made in another semantic domain. A non-legal semanticist would be likely to treat the verb *own* as a simple predicate (though possibly not a primitive), the functional noun *owner* as more complex, and the abstract noun *ownership* as being more complex still. A lexicologist of legal language, on the the other hand, is more likely to regard the concept of legal right as a primary notion in a network of definitions, and would therefore be more likely to define *ownership* as "the right (of a person) to use and dispose of (property)", and then to define both *owner* and *own* derivatively: an *owner* is a person who has such a right, *to own* is to have the right of ownership over some property.

Obviously the structuring of a vocabulary in terms of relative definitional complexity is going to vary a great deal depending on which of the many possible choices of a defining base the analyst decides to make.

Another methodological commitment to which an unyielding commitment would produce unprofitable results is what I have termed, in Fillmore (1975a), the "checklist theory of semantics". I have opposed checklist theories to theories of semantics which reorganize prototypes or paradigm cases for a category and which are unembarrassed by the occurrence, or even the prevalence, of word uses whose interpretations represent departures from the prototype. The checklist theory is one which seeks a list of criteria which state necessary and sufficient conditions for something to count as an instance of a category. The prototype theory is one which identifies the prototype case for a category, but which then allows something to be an instance of the category even if some of the identifying criteria are missing, or are only partially present.

To repeat an example that doesn't work as well as I would like it to work, I have proposed that the English verb *climb* has a "clambering" component, involving locomotory action of movable body parts, and an "ascending" component. A monkey climbing up a flagpole satisfies both of these conditions. The monkey climbing down the flagpole satisfies the clambering component only, but is nevertheless engaged in an action that can be properly called climbing. A snail climbing up the flagpole satisfies the "ascending" condition and can still be said to be climbing. But the snail is not privileged to *climb down* the flagpole, since that activity would involve neither clambering nor ascending. The prototype has two criteria, but either one of them can be missing.

Paul Kay and Linda Coleman (Kay and Coleman, to appear) have studied native speaker judgments on the category of telling a lie. The criteria for the prototype instance of *lie* include falsity of the spoken utterance, the speaker's awareness of the falsity of the spoken utterance, and the speaker's intention to deceive the hearer. By constructing situations in which each of these conditions was present or missing independently, and collecting subjects' graded judgments on which of these eight situations provided instances of lying, Kay and Coleman have discovered that most people accept as a lie situations in which any one of the three criteria is missing, many people accept as a lie a situation in which only one is present. (All of the constructed situations, of course, have in common the one necessary criterion of a lie, namely the property that one person is communicating a statement to another person.)

The difference, then, is that the prototype semantics makes it possible to talk about a central or nuclear sense of a word, and then, if necessary, about the various weightings of the individual criteria that go into specifying the prototype. The checklist semantics on the other hand is embarrased by fuzziness and degrees of category membership.

There are two additional related possibly dangerous methods, logical stances that I would like to discuss. The one is the decision to limit the semantic specification of lexical items to metalinguistic terms, that is, to the terms of a theoretical semantic descriptive system. The other, opposing, one is the decision to maximize cross-reference in a lexicon–that is, the decision, wherever possible, to show semantic relations between words by using defined words in definitions.

If you took the choice of maximizing cross-reference, then the lexicon, as a written document, would show certain kinds of conceptual prerequisite chaining relations between words in a fairly straightforward way. You look up a word, you look up the defined words contained in its definition, and so on until you have to stop. For English, *see* will be defined in some way; *look* might be defined as "act so as to *see*"; *stare* might be defined as "*look* sustainedly" or the like; and *gawk* will be defined as "*stare* while failing to hold one's mouth shut." *Stare* is in the definition of *gawk*, *look* is in the definition of *stare*, and *see* is in the definition of *look*.

A lexicon whose semantic specifications are restricted to as abstract meta-language could show the same sorts of relationships, of course, though maybe not so handily for the curious. But there might be facts that give us theoretically better reasons for allowing cross-reference in a lexicon. Robert Binnick (1971) has made a good argument for the analysis of *bring* as "cause to *come*" where the *come* in the definition is not the semantic primitive but the English word. Binnick's argument, roughly, is that otherwise the coincidence that so many of the uses of *come* are matched by analogous uses of *bring* could not be explained. Since you have to describe the whole coılection of idioms with *come* anyway, you do not have to mention those same facts all over again if you use the word *come* in the definition of the word *bring*.

In the same vein, Paul Kay and I have argued, in our course on lexical semantics that a polysemy in the word *day* can be used to explain a related polysemy in the word *morning*, by simply defining *morning* as the part of a day before noon. In both cases the word can refer either to a time unit that begins at midnight or one that begins at dawn.

I have noticed that the verbs *determine* and *price* have similar polysemy structures. *Determine* can mean either "decide effectively" or "find out". The question

> Have you determined when the concert will start?

can mean either of the following:

> Have you made up your mind about when the concert will start?

> Have you found out when the concert is going to

Analogously a statement like

> I spent the morning pricing boots.

can mean either of the following:

> I spent the morning deciding how much to charge for boots.

> I spent the morning learning how much boots cost.

This partial polysemy structure can be accounted for by defining *price* as "to determine the price of" and by allowing the polysemy of "price" to ride on the polysemy of "determine".[3]

An obvious fault of the cross-reference maximizing strategy is that it leaves us with a great number of arbitrary decisions to make. There are many inter-definable word sets in which any one word in the set could be used in the definition of each of the others. Examples are *up/down*, *right/left*, *start/stop/continue* and *permit/require/forbid*. In each case we have apparently arbitrary decisions to face, of the form "Shall we say that *hollow* means 'having a hole' or shall we define *hole* as 'a hollow place'?[4]

It seems to me that a commitment to either of these positions would be wrong, and that we need to be open to the possibility of lexical descriptions which would welcome both kinds of notation.

4.

A natural project for scholars interested in lexical semantics is that of characterizing patterned differences in lexical semantic structures across languages. Stephen Ullmann (1952, 1958), Ernst Leisi (1952), and many others, have made important contributions to such a project. Languages can be said to differ from each other, with respect to word semantics, along such dimensions as the following:

4.1. *Relative Analyticity*

This is what Ullmann refers to (1958, p. 680) as the dosage of arbitrary vs. motivated words in a language. One language might systematically have large repertories of unanalyzed words with highly specific meanings; another might have elaborate systems of semantically motivated word-formation principles; and a third might depend more on syntactic processes for expressing complex notions. Languages can be seen to differ strikingly in the relative weighting that is given to one or another of these devices.

---

3. *Bring* and *morning* offer stronger cases for cross-referencing than does *price*. The latter fits more naturally into an argument favoring semantic redundancy principles of the sort discussed by Gene Gragg (1978), since there are several others words–e.g., *establish*–which share the same polysemy pattern, making it impossible to decide nonarbitrarily which word should be taken as the one to which the polysemy is to be assigned.

4. *The Concise Oxford Dictionary*, as pointed out in Leisi (1952, p. 40), was unable to make a decision in the case of *hole* and *hollow*.

One of the relevant subdimensions here is the degree to which a language borrows from elsewhere its technical vocabulary. For people who do not know the source language, these imports come as unanalyzed words, whether or not they have that status in the source language. In English, "pain reliever" is motivated, "analgesic" is arbitrary.

By contrast to English, which has borrowed not only alien words but alien word-formation processes, languages that deliberately limit word formation to materials from the native stock, such as modern Hebrew and modern Icelandic, and to a large extent, standard German, have a higher proportion of motivated or semantically transparent words.

4.2. *Taxonomic Depth*

This is what Ullmann refers to as the dosage of particular and generic terms. In recent work this dimension has been studied most carefully, I believe, by Brent Berlin and his colleagues in connection with ethnobiological classifications (Berlin, Breedlove and Raven, 1973). Cultures whose members have achieved significant abilities to control nature are likely to have greater taxonomic depth to their plant and animal terminologies than others, and this both in the direction of greater abstractness or generality and in the direction of greater detail or specificity. Cultures that nurture legal institutions likewise invite extensive growth of taxonomies in the vocabulary of crime and punishment.

4.3. *Patterns of Meaning Extension*

Here I have in mind the systematic sorts of polysemy relations in particular domains of words. Languages may differ in the kinds of metaphoric extensions they have evolved, e.g., in their use of body-part names for general spatial concepts, in their extension of adjectives designating emotions to the causes of those emotions, etc. Some such developments depend on semantic properties as such and can be thought of as the language's acquisition of new semantic redundancy rules, as with the metaphoric extensions of color terms studied by Sandra Derrig (1978). Others seem to depend on real-world associations, such as the use, in Quebec French, for the most powerful and heavily emotional curses, of the vocabulary of the Christian church, as discussed in a recent study by Gillian Sankoff and Johanne Thibault. In this case the vocabulary used for cursing goes beyond the familiar *Christ* and *damn* but includes also such words as *chalice, host* and *tabernacle* (Sankoff and Thibault 1978).

4.4. *Areas of Synonymic Elaboration*

Here I have in mind synonymic elaborations of particular semantic domains where the differences serve certain social or pragmatic functions. English has, to give the obvious examples, alternate locutions for tabooed body parts, body products and bodily activities, these differentiated along such lines a Neutral, Clinical, Vulgar, Playful, etc. Other languages might ignore these domains and choose others for such vocabulary elaboration.

4.5. *Division of Labor Between Grammar and Lexicon*

Sometimes languages can differ from each other systematically in the patterns of assigning semantic information to the lexical or to the grammatical system. English sometimes uses aspectual information for discriminating lexical verbs from each other in ways that other languages would distinguish by grammatical means. Uto-Aztecan has many lexical verbs distinguished according to the singularity or plurality of their subjects; other languages either allow themselves to ignore such information or represent number inflectionally.

4.6. *Areas of Nomenclatural Specialization*

Here I have in mind domains in which more lexical discriminations are made in some languages than in others. The well-known examples are the Eskimo's snow words, the Bedouin's camel terminology, and the vocabulary for spiritual experiences in the languages of south Asia and in some dialects of American English spoken in northern California.

4.7. *Characteristic Lexicalization Patterns*

This is what Ullmann refers to as "characteristic tendencies". I have in mind the patterns by which different languages make such choices as lexicalizing motion with manner of motion, motion with path of motion, motion with the nature of the moving figure, etc. Some particularly insightful contributions to this study have been made by Leonard Talmy (esp. Talmy, 1972).

4.8. *Obligatory Categories*

Languages differ, not only grammatically but also lexically, in the sorts of obligatory categories they provide. We become most conscious of such phenomena when we find that our language has backed us into a corner, where whatever lexical choice we make brings with it presuppositions we may hot wish to subscribe to. The classical examples of what I have in mind are found in the speech level phenomena found in various parts of East Asia.

Closer to home are the conflicts sensed by anyone who takes seriously the effort to desex English. Since the system of pronouns in English is a closed class of words in which singularity for humans cannot be separated from sex, there is no way of choosing an anaphoric pronoun for an indefinite human antecedent without offending somebody. *They* offends the grammarians, *he* offends the feminists, *he or she* offends the stylists, *she* is downright hostile, and *it* just can't be taken seriously. We could get out of this by speaking Chinese, but that's bound to offend some people too.

4.9. *Collocational Patterns*

Languages can differ, lastly, in the degree to which portions of their lexica are collocationally bound. English has many words restricted to very special contexts–such as *blithering, excruciating,* and *kith*–while other languages may not be so endowed. It is difficult, of course, to know how to make reasonable estimates of the magnitude of such collocational binding.

5.

Probably in every language it is possible to discover and isolate what might be termed the general or primary vocabulary–roughly the sort of word set that corresponds to "Basic English", the vocabulary that is most likely to be intertranslatable across languages, the part of the vocabulary that, in the Queensland Australian situation described by Robert Dixon, is directly translatable into the Dyangul or "mother-in-law language" (Dixon, 1971).

In addition to the general vocabulary we find, at least in many languages, a considerable amount of vocabulary stratification. As I have already mentioned, languages that take in foreign words or foreign word-formation principles for their vocabulary of technology, religion or esthetics, have such stratification.

A phenomenon that might be particularly characteristic of such languages is the existence of large numbers of what might be called 'hard' words, words which only certain people know. I am not sure, but I would guess that not every linguistic community would find intelligible the desire to increase one's vocabulary that in our culture can exist quite independently of any desire to learn the skills or traditions which embody that vocabulary.

In one of those how-to-improve-your-vocabulary handbooks, I recently learned that I am a *misocapnist.* With a little bit of memory strain I might have been able to figure out, had I come across the word, that a misocapnist is somebody who doesn't like smoke; the definition I learned was 'one who hates the smell of tobacco smoke'. Now that I know the word, I can't think of what to do with it, unless I get a chance to design a button or bumper sticker that says, *Misocapnist Power.* My point is, of course, that languages can differ in including or not including in their word-hoard lexical items which are so little needed and which are likely to be learned only in a word-learning setting.)

More functional strata of words not known by everybody are those limited to association with particular trades, skills and cults. It is not at all uncommon for a culture to have large blocks of vocabulary that are known only to men, only to women, only to sorcerers or carpenters, or what have you.

Distinct from the situation in which vocabulary sets are only known to particular groups of people, we very frequently have situations in which words which are familiar to everybody are nevertheless only used by certain persons or in certain situations. Tony Woodbury's ongoing study of police talk shows extremely specialized kinds of vocabulary selection in police officers' public discourse about criminal justice. ("Shortly prior to the time of the explosion three male white individuals were observed to abandon their vehicle in this vicinity and to proceed northward hastily on foot.") In describing the ability of those who have mastered this register we will want to say that they have a controlled access to a somehow well-marked stratum of the vocabulary, together with a system of collocational preferences that have been developed for just this purpose.

6.

Words can differ from each other, as we have seen, in their relative semantic transparency, their 'motivation', their 'analyticity'. In a sense the arbitrary words are 'more difficult', since they have to be learned separately, whereas the motivated words are 'less difficult', because, even if the total meaning of the word is is not 'computable' from its parts, enough may be for the enhance ment of retention and recognition.

Chinese month-names are motivated, since they have the form '1-month', '2-month', etc. The English month-names, by contrast, are arbitrary–if you ignore the irrelevant etymological motivation of September, October, November, and December. Since these words are so thoroughly learned by every speaker of English, it may not be obvious that they could be 'difficult' words from the point of view of the scale of relative analyticity. But the effect of the difference between English and Chinese can be sensed if you can imagine an achievement test administered to American children in which the task is to associate a month-number with a month-name: on hearing June, the child says "6"; on hearing October, "10", etc., and if you can imagine that such a test successfully sorts children for some worthy educational propose. Now imagine that somebody gets the idea of translating this test into Chinese so that the same educational purposes could be served for Chinese children. The test would fail in this case, however, since the Chinese children would never make mistakes!

Month names in Saulteaux, according to A. Irving Hallowell (1955, pp. 226–227), are also motivated. The lunar months in the Saulteaux month-naming system are named according to phenomena in nature that occur during them. Thus, for example, there is the wild-berries-gathering moon, the blossom moon, the wild-rice-gathering moon, the rutting moon, the early winter moon, and so on. The Saulteaux, however, have not designated any of the moons in their cycle as the first one, and so there is no associated numbering from one to twelve with these words.

With the month names, in short, we have been able to compare three systems, one in which the names are arbitrary and two in which they are motivated, but according to different principles.

Differences in the degree and location of motivation can be brought out most sharply by comparing semantic systems which are clearly identical at the semantic level. Possibly the only one that works perfectly is the system of counting. Chinese, and hence Japanese, have high analyticity in their number systems. The 'teens' are 'ten-one', 'ten-two', etc., and the 'tys' are 'two-ten', 'three-ten' etc., English, by contrast, has more arbitrary elements–*eleven* and *twelve*–and a number of irregular or partially arbitrary elements, such as "teen", "-ty", and the variants of *two*, *three* and *five* found in *twenty*, *thirty* and *thirteen*, *fifty*, and *fifteen*.

In the large numbers, English has *hundred*, *thousand* and *million* as unmotivated terms, *billion*, *trillion* and *quadrillion* and the rest as motivated, at least to those who know the latinate system. In Chinese and Japanese the higher arbitrary numbers are *hundred*, *thousand* and *ten thousand*.

One of the clearest cases of cognitive difficulties related to semantic structural differences between languages can be seen in the low success with which speakers of European languages acquire mastery of the large numbers in Chinese and Japanese. The culprit is the monomorphemic name for 10,000.

In the case of the Chinese and English number systems, the critical difference comes with the location of higher-level arbitrary units such as the *ten-*

thousand morpheme of Chinese and the *million* morpheme of English. In this case, the difference is in the location, in the system, of these arbitrary elements. Each system is equally 'easy', I would guess, for its native speakers; and people brought up on each system will find difficulty in mastering the other. In other cases we will find that a vocabulary domain is simply more difficult in one language than in another because of the amount of motivation in these domains in the two languages. I suspect that German medical students have an easier time of it than American medical students if only because the technical body-part terms–names of muscles, for example–tend to be arbitrary in English, motivated in German.

Sometimes we also find that languages differ systematically across large vocabulary domains with respect to analyticity, even in ordinary parts of the language, words which everybody knows and uses regularly. Japanese and English, for example, differ systematically and very extensively in their lexicalization patterns. In particular, Japanese makes use of a very large set of mimetic adverbs, 'gitaigo' in the Japanese tradition, that combine with basic verbs to produce meanings that tend to be realized in English as monomorphemic verbs.

I have been trying to explore the extent of these differences between the two languages. Following a method that should have been obvious from the start but which in fact I first heard suggested by James McCawley, I have been finding word groups in *Roget's Thesaurus*, looking these up in a large English-Japanese dictionary, and then collecting and sorting all of the translations that center around some key word in the Japanese. I performed this exercise, for example, with verbs in the general domain of *walking*.

The Japanese translation of *walk* is *aruku*, or, more accurately for most uses, *aruite iku* (go walking) or *aruite kuru* (come walking). For many of the English verbs in the walking domain, the Japanese translation consisted of the word *aruku* plus one of a number of adverbial elements. Some of these adverbial elements are phrasal, but others are instances of the class of manner adverbs I mentioned earlier. Examples of such manner adverbs coming out of the *walking* exercise are *burabura, yochiyochi, tobotobo, yoroyoro* and *doshindoshinto. Stroll, saunter* and *meander* all yielded *burabura aruku*; *plod* was *tobotobo aruku*; *totter* was *yochiyochi aruku*; *waddle* was *yoroyoro aruku*; and *lumber* was *doshindoshinto aruku.*

Now it happens that some of these adverbs are only used connections with styles of walking, so that although *burabura aruku* is 'analytic' with respect to *aruku* (walk), the adverb itself is arbitrary and limited to this semantic context. (Thus, as in English one learns non-analytically to distinguish *waddle* from *walk*, in Japanese, analogously, one learns non-analytically to distinguish *yoroyoro aruku* from simple *aruku.*) *Doshindoshinto*, the one that occurred in the translation of *lumber*, is an onomatopoetic word; *doshindoshin* had *thump* as its translation.

In a number of other cases the translations were more analytic. Thus, the translation of *stride* meant 'to walk with big steps'; *trudge* and *plod* were both described as 'walking in a weighed down or burdened manner' (*omokkurushiku aruku*); *swagger* was 'to walk being proud' (*ibatte aruku*); and *strut* was 'to walk being affected' (*kidotte aruku*).

Now many of the more analytic translations are not properly speaking translations but are rather explanations of the corresponding English words. What I mean by that is that these adverbials wouldn't ordinarily be used in situations analogous to those in which such English verbs as *stride*, *strut* and *swagger* would be used.

The lexical semantic structural observations that come out of this, in any case, are that in English we have to do with a block of morphologically simple verbs having 'walk' as their semantic core and distinguished from each other in a wide variety of ways having to do with manner and attitude, and that in Japanese we have a single verb that means 'walk' and a cluster of adverbs that seem to be specially evolved to depict manners of walking. The more analytic translations do not figure in a study of lexical structure, on the Japanese end, since these are constructed according to the usual kinds of syntactic phrase-forming processes.

7.

I would like next to consider the ways in which different semantic properties of a complex notion get differently realized morphologically. I would like to illustrate this concern by considering verbs that express complex events containing the notion of Result. That is, I will be concerned here with events thought of as made up of two parts, the first being an event or act, the second being an event or situation which is seen as resulting from the first one. I would like to explore the kind of morphological relations that can exist between the verb that names the initial event or the resulting event and the verb that names the whole event.

The possibilities can be presented as follows: let X stand for the verb which names the initial event; let Y stand for a verb or adjective describing the resulting situation; let BECOME Y stand for the event of the resulting situation coming about; and let either X CAUSE Y (where Y is an event verb) or X

CAUSE BECOME Y (where Y is a state word) stand for the verbal part of the descriptions of the total events. I would like to use the symbol Z for a single verb which expresses the total complex event.

One possibility is for there to be no encompassing verb Z. Thus, if you do something that makes me squeal, there is no single verb which describes that whole event. I cannot say, for example, that you *besquealed* me, meaning just that.

Another possibility is for the verb Z to be unrelated morphologically to either X or Y. Thus, *kill* is the verb which names a situation in which one entity is involved in an event with the result that something dies. Here the X is an unspecified verb and Y is *die*, Z being unanalyzable with respect to either X or Y. A possibility we find a great deal in English is for the verb Z to be the same as the verb Y. If by doing something the result is that something moves, we can say that we *moved* that thing. Y and Z are expressible by means of the same verb.

Another possibility in English is for the verb Z to be the same as the verb X. If by pushing against something I cause it to move, we can say that I *pushed* it. Here the match is not perfect: the original event is best described with "push against" and the total event with just "push"; but we can still regard this as a case where the same morphological material is used for X and Z.

Another possibility, somewhat more restricted in English, is that in which Z is a derivative of X. Thus, if by smearing something on a thing I cause that thing to be covered by the substance smeared, we can say that I *besmeared* it.

A final possibility, found only in phrasal verbs in English, is that in which Z is morphologically X plus Y. The Mandarin resultative construction gives us examples of this situation, one being the verb analytically translated as "cry-red". If by crying I can cause my eyes to become red, I can say *I cry-red eye* to mean just that. There is no transitive verb meaning *break*, but there's an XY word meaning *hit-break* which serves that function.

I have mentioned all of the possibilities because it turns out that languages differ sometimes systematically in their use of these devices. English is rare in having so many cases in which Z is identical in form with X and Y. Hungarian and German have regular ways of forming Z as a derivative of X. Mandarin has many words constructed in the X-Y form.

The point here has been that languages can differ systematically with respect to how they provide morphological realization of kinds of semantic patterns. The character of this relationship is one aspect of lexical structure.

8.

Possibly the most extensively discussed kind of lexical structure is that of the various word-fields or semantic domains. Word field studies take the vocabulary of some particular domain–color, kinship, perception, temperature, body-part names, etc.–and classify the semantic features which serve to distinguish the words in such a set from each other.

The English verbs of motion have been given extensive treatment by George Miller (1972), Dorothea Weniger (1974), Yasuhiko Ikegami (1970), and many others. The words in such a set are typically morphologically unanalyzed. In the case of the motion verbs, the relevant dimensions can be classified in an unlimited number of ways, but including at least 'path of movement' (as with *ascend* and *advance*), 'path of movement in connection with some ground' (as *climb, dive, cross*); 'path of movement with respect to the starting point or the destination of the movement' (as with *arrive, alight, leave, enter*); 'manner of movement' (as with *lope, stride, scurry, slog*); 'noise accompanying the movement' (as with *swish, stamp, scuffle*); 'body involvement' (as with *stride, creep, crawl*); 'speed of movement' (as with *bolt, dart, hurry*); and so on and on.

Body part semantics, to consider another domain, offers its own special set of questions, as outlined in a recent paper by Elaine Anderson (Anderson 1977). Here one can ask, for example, which words in the domain are specialized for human body parts, as opposed to animal body parts in general; which are primarily nonbody-part terms that have been extended into the body-part domains; which human body-part names have been extended outward into other domains; which parts of the body have names; what is the morphological structure of these names; and so on.

Instead of examining specific domains as such, what I would like to do here is speak about the various kinds of structures that characterize different semantic domains. One structural notion, that of markedness, figures in several of these domain structures and will be mentioned whenever it appears to be relevant.

First of all, as the simplest kind of domain structure, we have the Contrast Set. Some words come in obvious and usually small associative sets, it being the case that to know the meaning of one member of a contrast set is to understand the whole set. The members of an 'opposition' are not knowable independently of the opposition. The simplest cases are the various forms of semantic opposition–'tall'/'short', 'girl'/'woman', 'dead'/'alive', 'raw'/'cooked', etc. For some contrast sets the question of markedness comes up just in case one of the terms can be used, in some contexts, as a names for the oppositional category as a whole.

Another kind of semantic domain structure is that of a Taxonomy. A taxonomy is defined as a set of words linked to each other through the relation of superordination. In a taxonomy the markedness phenomenon takes the form of a single word which can serve as both a co-hyponym and a superordinate of some other term, as for example in a taxonomy in which the word meaning 'oak' is identical with the word meaning 'tree'.

Recent work has suggested that taxonomies are more than just sets of words linked by the superordinate relation; that there is, in fact, a special privileged status, at least for many taxonomies, according to a particular level in a taxonomy, such that higher-level taxa and lower-level subcategories alike have to be understood as existing for special reasons, both in the evolution of the taxonomy and in the synchronic functioning of the taxonomy. I have in mind, of course, the work of Eleanor Rosch on the notion of basic level categories (see Rosch 1973) and the work of Brent Berlin and his colleagues on the notion of generic categories (Berlin, Breedlove and Raven 1973).

Another kind of domain structure in semantics is that of the Partonomy. A partonomy is a collections of words linked by a part-whole relationship. Body part terms provide the most obviously structured partonomies (a fingernail is a part of a finger, a finger is a part of a hand, a hand is a part of an arm, an arm is a part of a body), but many others are possible. In a partonomy the marked/unmarked contrast figures in the case of words which name both a part and its immediate 'whole' (as in languages which have the same word for foot and leg) and in the case of words which name analogous 'parts' of different 'wholes' (as in languages which have the same words for fingers and toes).

Quasi-partonomies exist in which the part of relation is not transitive; these, in fact, quite probably predominate. Is is possible to say that a fingernail is part of a finger, maybe all right to say that a fingernail is part of a hand, wrong (or maybe just uninformative?) to say that a fingernail is part of an arm; but possibly all right again to say that a fingernail is part of the body.

The best known and most discussed kind of semantic domain structure is the Paradigm. A paradigm is a group of words having one feature in common and distinguished from each other by features that each discriminate more than one pair of words. (The distinction between *wolf* and *fox* is not paradigmatic because it does not distinguish any other pair of words; the distinction between *man* and *woman* is paradigmatic because it also figures in the distinction between *boy* and *girl*, *actor* and *actress* and a great many others.) A perfect paradigm is one in which every possible combination of features is represented by one preferably monomorphemic word.

There do not appear to be many large paradigms around, at least not in the general vocabulary. A standard one in the handbooks is the livestock paradigm, in which words having to do with farm animals are arrayed according to such features as 'male'/'female'/'neutered', 'adult'/'young'/'infant', 'name of caretaker', 'name for bearing offspring', 'name for copulation', and so on. There are gaps in the paradigm–there not seeming to be any word for a young female sheep that would correspond to *heifer, filly* and *gilt*; and the paradigm cannot be said to be a particularly active one in the usage of most native speakers of English, since such words as *wether, gilt, barrow, whelp* or *sup* are not actually known to very many speakers of the language.

In studying paradigms it is valuable to discover gaps–such as that of the missing item for young female sheep–and to notice what productive devices are used for filling them. In comparing languages it is important to notice whether the same features have paradigmatic roles in both languages. Bakweri, a Bantu language of the Cameroun, seems to have no use in its lexicon for a paradigmatic distinction between adult and child. 'Boy' is 'child man', 'girl' is 'child woman', 'puppy' is 'child dog', and so on.

Another minor domain structure is the Cycle. Examples of cycles are "morning"/"afternoon"/"evening"/"night" and "spring"/"summer"/"fall"/ "winter". Some cycles represent natural phenomena, others, such as the weekday names, stand for artificial cyclic systems.

Another minor domain structure is the Chain. A chain is a set of words linked by some sort of rank ordering relation and differing from a partonomy or a taxonomy in not allowing the possibility of two terms at the same level. (Partonomies and taxonomies, in other words, can be represented as branching tree diagrams; there is no 'branching' in a chain.) Military titles provide an example of a chain-like structure; the names of the speeds on your electric blender provide another.

A still different kind of semantic domain structure is what might be called the Network, illustrated by systems of kinship terminology. A network is constructed by a small number of primitive relations (rather than by a single relation, as with a partonomy or a taxonomy). In the case of kinship terminology these are most typically 'married to' and 'parent of', with 'older than' a common addition. The way to study a semantic network is to notice which relationships definable in the network are named, which are not, which patterns of relationships have the same names, what notions other than the basic relations defining the network get encoded into the system of terms (sex, social status, etc.), and, of course, how such structurings differ across languages.

A final kind of semantic domain type–which I happen to think is the most central and powerful of them all–is what we might call, not too helpfully, a Frame. A 'frame' in this sense is a lexical set whose members index portions or aspects of some conceptual or actional whole. The items in a frame, in other words, are only understandable to somebody who has (conceptual) access to the underlying schema onto which the parts of the frame fit. The notion is easier to exemplify than to explain: one example is that found in connection with the 'commercial event'. The event type is one in which one person exchanges money for some sort of goods or services received from a second person. There is a large set of words that key onto various parts and aspects of the commercial event schema. Examples are *buy, sell, pay, spend, cost, charge, price, money, change*, and dozens of others. Within the set of words linked together in a 'frame' can be found many that form paradigms, contrast sets, taxonomies, and the rest; but all of them require, for their semantic specification, a prior detailing of the nature of the associated conceptual schema.

9.

I think that semantic theory must reject the suggestion that all meanings need to be described in the same terms. I think, in fact, that semantic domains are going to differ from each other according to the kind of 'definitional base' which is most appropriate to them. I think there will be some domains in which binary oppositions will be fitting and necessary; others where it will not. I think there are some domains in which it will make sense to look for the most primitive underlying semantic building blocks, others in which it will not; some in which appeals to knowledge about objects and institutions in the world will be necessary, and others where they will not.

There are, in short, various kinds of word meanings. Some words seem to have a more or less straightforwardly statable relationship with some aspect of the world–such as, for example, the color words, or the names of natural kinds, such as uncomplicated and familiar plants and animals. Other words (*pilot, diary, teacup*) have to be described as designating objects by reference to their function or their characteristic activity. Still others (*lazy, witty, genius, fool*) designate entities according to their dependable dispositions. Still others (*widow, tenure, apple core*) can be understood only within the framework of an understanding of a social institution or a way of doing things. For many words of this latter kind, they can be said to have lexical structure only by virtue of the structuring of the non-linguistic activities or institutions whose parts and aspects they index.

Accordingly, and in contrast with what I once claimed (Fillmore 1975, pp.

143–146) I think that all of the various kinds of definitions have a role to play in linguistic semantics.

Scholars have distinguished: (A) 'Stipulating definitions', by which we recognize that the relation between an expression and its meaning exists because some actual person performed a defining illocutionary act. (In linguistics the definition of *transformation* is surely like that.) (B) 'Characterizing definitions', with which the definers attempt to capture the essence of a category that is understood independently of knowing the definition of the word which names it. (As with linguists' frequent attempts to define *word* or *language.*) (C) 'Criterial definitions', by which an attempt is made to discover some distinguishing mark of the category, whether or not this can be taken as presenting the essence of the associated category or the use conditions of the associated word. (An example is the occasional linguistic definition of *verb* in English as a word capable of taking the progressive suffix.) And lastly, (D), what might be called a 'use description definition', by which the definer tries to determine the conditions governing observed uses of a word, as seen, for example, in linguists' attempts to define, say, a word like *remind.* The checklist variety of this sort of definition is the attempt to provide the necessary and sufficient conditions for membership in the category designated by the word; the prototype version is one which recognizes the fuzziness of a category and which distinguishes a prototype use of a word from its more peripheral uses.

10.

Having set up all this conceptual apparatus, I would now like to turn to an examination of legal language and the special semantic theories that seem to be assumed by those who use it. I do this, apart from whatever is inherently interesting about legal language, for the sake of setting up a standard of contrast for some generalizations about the semantics of ordinary language.[5]

It doesn't take long to recognize that the language of law has its own semantic guidelines. I think that some of the differences can be apparent if we listen in on a conversation that took place in a voir dire examination in a criminal court in Berkeley, California, about a year ago.

---

5. I should explain that I do not have, and do not claim, expertise in the law. My knowledge of legal language comes out of two hobbies, one that of spending time tracing my way through dictionaries, including law dictionaries, the other that of parsing sentences taken from legal texts. Although I have checked some of my remarks here with people who know more about the law than I do–Patrick Cathcart and Ronya Robinson–there still might be serious misrepresentations in what I say about legal language.

The examining attorney asked the prospective juror the following question:

> Do you accept without question the American legal doctrine that a man is innocent until proven guilty?

The prospective juror responded:

> I think a man should be *treated* as innocent until proven guilty, but it would be funny to say that he *is* innocent until proven guilty.

The interrogator:

> The doctrine I am speaking of is precisely that a man *is* innocent until proven guilty. Do you or do you not accept that doctrine?

The prospective juror:

> But if he *is* innocent, then it would be *wrong* to prove him guilty. So why have the trial?

The prospective juror was challenged for cause and excused from jury service.

It is important for a juror to understand that the proper judgment is *not guilty* for anybody who has not been proven guilty. The departure from the ordinary way of talking is that the presumption-of-innocence doctrine should be expressed by the statement that the person *is* innocent.

The law, then, has its own sort of semantic principles. I mentioned earlier that the basis of definitions is frequently one of 'rights' and 'responsibilities', so that, to repeat the example, *ownership*, while morphologically more complex than *owner* is semantically simpler: an *owner* is one who has the right of *ownership*.

The semantics of the technical stratum of legal language is founded on stipulative definitions. In a sense, legal language shows us a checklist semantics gone mad. At the same time, of course, there is a great deal of vagueness deliberately built in—as with such concepts as *intention*, *malice*, *clear and present danger*, and so on.

Legal language has achieved an effective compromise between prototype semantics and checklist semantics, this in the form of various principles of statutory interpretation. A prototype definition of *murder* involves killing somebody, intending to kill the person that got killed, and having a malignant heart about the whole thing. The really clear cases of murder will include all of those. One kind of departure from the prototype is found in a situation in which, say, wishing to kill Smith, having a malignant heart about killing Smith, aiming your gun at Smith and pulling the trigger, you nevertheless succeed only in killing Jones who was standing by Smith's side. With this mistake you have divided the conditions for committing murder between Smith and Jones. Was what you did an act of murder?

The official answer comes by appeal to the doctrine of Transfer of Intent– a principle which covers just such cases. The prototype definition of murder stands unamended; a fuzziness removing principle is invoked to make the departure from the prototype count as a full instance of the category. The principle is that if while intending to commit a criminal act you instead do something that would have counted as a criminal act if you had intended to do that, then the law says you intended to do that second act.

The language of criminal justice is filled with contrast sets whose discriminating features are the result of stipulating definitions. The difference between *jail* and *prison* in many states is determined by the duration of incarceration; if it's a place for incarcerating folks for less than one year, it's a jail; a year or more, it's a prison. In some jurisdictions, according to Ronya Robinson, the difference between simple assault and aggravated assault depends on whether or not the victim had to be hospitalized more than 72 hours. (You may need to wait three days before you find out what crime you committed.)

Legal language has a great many paradigms and partial paradigms. What distinguished *citation*, *summons* and *subpoena* is determined by who receives the document and who authorizes its delivery. In the case of *citation* and *summons*, the receiver is the accused; with *subpoena* it is the witness; in the case of *summons* and *subpoena* the authorizing agency is a judicial officer, with a *citation*, a law enforcement officer.

The language of law is filled with taxonomies. Some simple taxonomies make use of discriminations that have been operationalized and simple nested contrast sets. There are three kinds of *offenses*: *infractions*, *misdemeanors*, and *felonies*. *Misdemeanors* and *felonies*, as opposed to *infractions*, are punishable by incarceration; and they are separated from each other by whether the incarceration is in jail or in prison.

Homicide is a high-level taxon having three subparts, called *excusable*, *justifiable* and *felonious*. Felonious homicide has three subcategories: *suicide*, *manslaughter* and *murder*. And *murder*, in turn, is subcategorized according to degrees of seriousness.

I would like to look in some detail at one block of legal terminology. The language of inheritance law contains contrast sets, paradigms, networks, taxonomies, frames, the works. Property is divided into the inheritable and the noninheritable, the former referred to as *hereditaments*. Hereditaments can be *movable* or *immovable*, the actual criterion for this distinction, according to one ruling, being whether the thing can be moved without damaging something else. A tree is immovable because if you move it you leave a big hole; a bicycle is movable. Movable property is called *personal*; immovable property is called *real*. In the context of discussions of inheritance, personal property can be referred to as *personalty*, real property as *realty*.

A person wishing to determine how his or her property is inherited will draw up a will. Having done that he or she has become a *testator* or a *testatrix*. The usual will or testament is drawn up with the aid of an attorney. However, the law recognizes a *holographic will*, drawn up by hand, and a *nuncupative will*, made orally. A nuncupative will is an oral testament, spoken in the presence of two or more witnesses, concerning property valued at less than fifty dollars, by a military person in combat or by a merchant sailor while in waters subject to the ebb and flow of the tide.

Here is some more terminology. The state of dying having made a will is called *testacy*; the state of dying without having made a will is called *intestacy*. Realty that is inherited from someone who dies *testate* is called *devise*. Personalty inherited from someone who dies testate is called *legacy*. Realty inherited from someone who dies *intestate* is referred to as *descent*; personally inherited from someone who dies intestate is called *distribution*. The receiver of a devise is a *devisee*; the receiver of a legacy is a *legatee*; the receiver of a *descent* is called an *heir* and the receiver of a *distribution* is referred to as *next of kin*".[6] *Dower* is a wife's right to her husband's property. *Curtesy* is the name of a man's right to his wife's property, but in this case the word is fitting only if there has been *issue*. While his wife is living this right is called *curtesy initiate*, after her death it is called *curtesy consummate*.

The purpose of this display was to set up some data on the lexical semantic situation in legal language so that we can compare that situation with what we find in the language of the folk. In the special case of legal language, we know what unusual requirements the accepted principles of word interpretation must meet. In doing the comparison we can ask whether and where there are

6. It is difficult to become fluent in legal language when one's source is limited to dictionaries and dictionary-like handbooks. In particular, I have not been able to learn anything about the syntax of the nouns *descent* and *distribution*; for example, I do not know whether they are count nouns or mass nouns.

analogous requirements in ordinary language.

We can take either of two positions with respect to the comparison. We could claim that legal language is, after all, language, and argue that a theory of linguistic semantics should include principles for dealing with such systems; alternatively we could claim that legal language is an artificial language that has nothing to do with what linguists should spend their time worrying about. In either case we may find the comparison helpful.

In the first place, we have already noticed that the definitional base for words that are given legal definitions is frequently different from what might be proposed for the same words in ordinary language. This being so, say about situations in which the law has its own definitions of words that have non-technical uses as well. In the case of *ownership*, for example, shall we say that this is the same word as the word *ownership* we find in non-legal contexts? Or should we say that *ownership*, for all its being a familiar and everyday word, is a legal concept from the beginning and that the legal definition is simply the correct one? Or could we say that the same word, in the same meaning, can have different semantic descriptions, all of them correct, for different purposes?

In ordinary language a great many word meanings are left fuzzy, their application depending on cooperation and common sense, or simply left unresolved. We might be sure about clear instances of being *alive* and equally clear instances of being *dead*, but in ordinary life we have no need to redefine the notions so that the borderline becomes a precise one. The law, however, might indeed need to make the distinction a precise one, that is, to convert a prototype semantics into a checklist semantics. It is clear that, since on the question of death courts consult biologists and physicians rather than lexicographers, the statement that eventually gets stipulated is not arrived at by the same process that disputing linguists go through in trying to determine what in fact is the correct meaning description of some word.

Workers in lexical semantics sometimes raise the question of what is a possible lexical item (McCawley 1971). I suspect that such a question has a different meaning when posed of lexical items in the folk lexicon from what it might have in a legal or technical context. The serious question of whether disjunctions should be allowed in semantic specifications has a trivial answer in a system of terms evolved through generations of stipulation and re-interpretation. In legal language, disjunctive definitions can be found at every hand–as in the definition of *tort*, for example, or, as we saw, in the definition of *nuncupative will*.

Legal vocabulary, more than anything else I know of, forces us to recognize phrases as lexical items. I think that the words *ingress* and *egress* occur only in the phrase, *the right of ingress and egress*, which, as a whole phrase, is defined as 'the right of a lessee to enter and leave a leasehold'. The semantics is compositional and noncompositional at the same time. *Clear and present danger* is another such term, and there are hundreds more. Semantic theory clearly needs to recognize 'idioms', having a noncompositional semantic structure, 'phraseological units', phrasal units that are partly arbitrary and partly motivated, as well as non-idiomatic 'formulas', expressions whose frequency or popularity, rather than meaning, is conventional. Perhaps in the case of nonidiomatic phrases that need nevertheless to be defined as units, of the kind we find in the law, we are dealing with something distinct still; or perhaps there is a critical difference between semantic specification, stipulative or not, and principles of statutory interpretation.

Legal taxonomies expand somewhat in the ways pointed out by Berlin, Breedlove and Raven (1973), but in this case the reasons depend on institutional decisions rather than knowledge of biological facts; and, in this case, I would guess there is no natural upper limit to the possible 'depth' of a taxonomy. Further regroupings, upward in the direction of higher level generalizations, and further subdivisions, downward in the direction of greater specificity, seem always to be possible. In the taxonomy that was built around *murder*, the lay interpreter will still want to say that *murder* is the basic level category; but it is apparent that the concept of generic or basic-level category does not have a role within the semantic system of law.

Legal language is in many ways similar to and in many ways different from ordinary language. Its private vocabulary can be regarded as a separate stratum, some words belonging to this stratum because they have meanings that are specially defined for legal purposes, some because they exist only in the legal vocabulary, others because they have been chosen for the legal register on the grounds that they add precision–anaphoric *said* and cataphoric *hereinafter*–or because they add solemnity.

11.

A question I invited you to consider was whether the semantics of law was different in principle from the semantics of ordinary language. The position I hope I have led you to is the position that there is no such thing as a uniform semantics for ordinary language, and that therefore we ought to be interested in the language of law for its own sake. I think that a decision to limit oneself to some more central domain of the vocabulary of a language should be seen as a decision based on interest rather than on any assumptions about what is properly a part of the domain of linguistic science.

If this paper can be said to have a theme, it is mainly a negative one, not exactly anticomponentialist, but antiexclusivist. People need to categorize objects and events in their worlds. When we wish to study instances of categorization items in their language, we can do this only by asking what functions such categorizations have in their lives. Some of the categorizations we find have only linguistic explanations: people do it that way because that's how their language evolved, and it could have evolved in a number of other ways. Others have, at least in part, explanations that depend in crucial ways on such matters as how human beings perceive things in the world around them, how people form categorizations in general, and what social institutions form the matrix of their daily activities. We need theories of word meanings that will not encourage us to lose sight of these realities.

References

Abaev, V. I. (1964). "A grammatical sketch of Ossetic (Edited by Herbert H. Paper. Translated by Steven P. Hill)." *International Journal of American Linguistics, publication of Indiana University Research Center in Anthropology, Folklore, and Linguistics.*

Abelson, Robert (1973). The structure of belief systems. *Computer Models of Thought and Language.* Roger Schank and Kenneth M. Colby. San Francisco, Freeman.

Anderson, Elaine (1977). "The eyes have it: lexical structure and domain of human body parts." *Ms.*

Anderson, John (1971). *The Grammar of Case: Towards a Localistic Theory.* Cambridge, Cambridge University Press.

Anderson, Stephen R. (1971). "On the role of deep structure in semantic interpretation." *Foundations of Language* **6**: 197–219.

Austin, John L. (1965). *How to do Things with Words.* New York, Oxford University Press.

Babcock, Sandra S. (1966). "Syntactic dissimilation." *Unpublished. Ohio State University.*

Bach, Emmon (1965). On some recurrent types of transformations. *Report of the Sixteenth Annual Round Table Meeting on Linguistics and Language Studies (identical with Georgetown University Monograph Series on Languages and Linguistics 18).* Charles W. Kreidler. Washington, D.C., Georgetown University Press: 3–18.

Bach, Emmon (1967). "'Have' and 'be' in English syntax." *Language* **43**: 462–485.

Bach, Emmon (1968). Nouns and noun phrases. *Universals in Linguistic Theory.* E. Bach and R. Harms. New York, Holt, Rinehart and Winston.

Bally, Charles (1936). L'expression des idées de sphère personnelle et de solidarité dans les langues indoeuropéenes. *Festschrift Louis Gauchat.* Aarau: 68–78.

Baugh, Albert C. (1957). *A History of the English Language.* New York, Appleton-Century-Crofts.

Bazell, C. E. (1949). "Syntactic relations and linguistic typology." *Cahiers Ferdinand de Saussure* **8**: 5–20.

Bendix, Edward H. (1966). *Componential Analysis of General Vocabulary: The Semantic Structure of a Set of Verbs in English, Hindi and Japanese.* Bloomington, Indiana University Press.

Bennett, Charles (1914). *Syntax of Early Latin, II: The cases.* Boston, Allyn & Bacon.

Benveniste, Emile (1962). "Pour l'analyse des fonctions casuelles: Le génétif latin." *Lingua* **11**: 10–18.

Berlin, Brent, Dennis E. Breedlove & Peter H. Raven (1973). "General principles of classification and nomenclature in folk biology." *American Anthropologist* **75**: 214–242.

Bernhardi, August Ferdinand (1805). *Anfangsgrunde der Sprachwissenschaft.* Stuttgart, Frommann-Holzboog, 1990 (originally published: Berlin: H. Frölich, 1805).

Bierwisch, Manfred (1967). "Some semantic universals of German adjectivals." *Foundations of Language* **3**: 1–36.

Bierwisch, Manfred and Ferenc Kiefer (1966). Remarks on definitions in natural language. *Studies in Syntax and Semantics.* F. Kiefer. Dordrecht, Reidel.

Binnick, Robert (1971). "Bring and come." *Linguistic Inquiry* **2**: 260–265.

Blake, Frank (1930). A semantic analysis of case. *Curme volume of linguistic studies (identical with Language Monograph No. 7).* Werner Leopold James Taft Hartfield, and A. J. Friedrich Ziegelschmid. Baltimore, Waverly Press, Inc.: 34–49.

Cassidy, Frederick G. (1937). "'Case' in Modern English." *Language* **13**: 240–245.

Chafe, Wallace L. (1970). *Meaning and the Structure of Language.* Chicago, University of Chicago Press.

Chafe, Wallace L. (1974). "Language and consciousness." *Language* **50**(1): 111–133.

Charniak, Eugene (1974). He will make you take it back: a study in the pragmatics of language. Castagnola, Switzerland, Technical Report, Istituto per gli studi semantici e cognitivi.

Chomsky, Noam (1965). *Aspects of the Theory of Syntax.* Cambridge, M.I.T. Press.

Chomsky, Noam (1966). *Cartesian Linguistics.* New York, Harper & Row.

Chomsky, Noam (1970). Deep structure, surface structure, and semantic interpretation. *Studies in General and Oriental Linguistics.* R. Jakobs and S. Kawamoto. Tokyo, TEC Corporation for Language Research.

Coles, L. Stephen (1972). Techniques for Information Retrieval Using an Inferential Question-answering System with Natural-language Input. Stanford, Stanford Research Institute, Stanford University.

Cook, Walter J. (1972). A set of postulates for case grammar analysis. *Languages and Linguistics Working Papers No. 4.* Washington D. C., Georgetown University Press: 35–49.

Cruse, D. Allen (1973). "Some thoughts on agentivity." *Journal of Linguistics* **9**: 1–23.

de Groot, A. Willem (1956). "Classification of uses of a case illustrated on the genitive in Latin." *Lingua* **6**: 8–66.

Diver, William (1964). "The system of agency of the Latin noun." *Word* **20**: 178–196.

Dixon, Robert M. W. (1971). A method of semantic description. *Semantics: An Interdisciplinary Reader in Philosophy, Linguistics and Psychology.* Danny Steinberg and Leon Jakobovits. Cambridge, Cambridge University Press.

Dougherty, Ray C. (1974). "Generative semantic methods, a Bloomfieldian counterrevolution." *International Journal of Dravidian Linguistics* **3**: 255–286.

Emons, Rudolf (1974). *Valenzen Englischer Prädikatsverben.* Tübingen, Niemeyer.

Fillmore, Charles J. (1965). "Entailment Rules in a Semantic Theory." *Project on Linguistic Analysis* Report No. 10.

Fillmore, Charles J. (1966a). Toward a modern theory of case. *The Ohio State University project on linguistic analysis, Report No. 13.* Columbus, Ohio State University: 1–24.

Fillmore, Charles J. (1966b). "Deictic categories in the semantics of 'come'." *Foundations of Language* **2**: 219–227.

Fillmore, Charles J. (1967). "The syntax of English preverbs." *Glossa* **1**: 91–125.

Fillmore, Charles J. (1968). The case for case. *Universals in Linguistic Theory.* E. Bach and R. Harms. New York, Holt, Rinehart, and Winston: 1–89.

Fillmore, Charles J. (1972). On generativity. *The Goals of Linguistic Theory.* Stanley Peters. Englewood Cliffs, Prentice-Hall: 1–19.

Fillmore, Charles J. (1975a). An alternative to checklist theories of meaning. *Proceedings of the First Annual Meeting of the Berkeley Linguistic Society.* Catherine Cogen et al. Berkeley, Berkeley Linguistic Society: 123–131.

Fillmore, Charles J. (1975b). The future of semantics. *The Scope of American Linguistics.* Robert Austerlitz. Lisse, Peter de Ridder Press: 135–158.

Fillmore, Charles J. (1977). The case for case reopened. *Syntax and Semantics 8: Grammatical Relations.* Peter Cole and Jerrold Sadock. New York, Academic Press: 59–82.

Finke, Peter (1974). *Theoretische Probleme der Kasusgrammatik.* Kronberg, Scriptor Verlag.

Frei, Henri (1939). Sylvie est jolie des yeux. *Mélanges de linguistique offerts à Charles Bally.* s.a. Georg et cie. Geneva: 185–192.

Frei, Henri (1954). "Cas et dèsinences en francais." *Cahiers Ferdinand de Saussure* **12**: 29–47.

Gleitman, Lila (1965). "Coordinating conjunctions in English." *Language* **41**: 260–293.

Gonda, Jan (1962). "The unity of the Vedic dative." *Lingua* **11**: 141–150.

Greenberg, Joseph H. (1963). Some universals of grammar with particular reference to the order of meaningful elements. *Universals of Language*. Joseph H. Greenberg. Cambridge, M.I.T. Press: 58–90.

Greenberg, Joseph H. (1966). Language Universals. *Current Trends in Linguistics*. Thomas A. Sebeok. The Hague, Mouton. **III**: 61–112.

Grimes, Joseph E. (1964). *Huichol syntax*. The Hague, Mouton.

Gruber, Jeffrey S. (1965). "Studies in Lexical Relations." *M.I.T. Unpublished dissertation*: 24.

Gruber, Jeffrey S. (1967a). "Look and See." *Language* **43**(4): 937–947.

Gruber, Jeffrey S. (1967b). "Topicalization in child language." *Foundations of Language* **3**: 37–65.

Gruber, Jeffrey S. (1967c). Functions of the Lexicon in Formal Descriptive Grammars. *Technical Memorandum TM-3770/000/00*. Santa Monica, System Development Corporation.

Hall, Barbara (1965). "Subject and object in English." *Unpublished dissertation. M.I.T.*

Halliday, Michael A. K. (1966). "Some notes on 'deep' grammar." *Journal of Linguistics* **2**: 55–67.

Halliday, M. A. K. (1967). "Notes on transitivity and theme in English, Part 1." *Journal of Linguistics* **3**: 37–81.

Halliday, Michael A. K. (1970). Language structure and language function. *New Horizons in Linguistics*. John Lyons. London, Penguin Books: 140–165.

Hallowell, A. Irving (1955). *Culture and Experience*. Philadelphia, University of Pennsylvania Press (references here are from the 1971 Schocken paperback edition).

Harris, Zellig (1957). "Co-occurrence and transformation in linguistic structure." *Language* **33**: 283–340.

Hashimoto, Mantaro J. (1966). *The internal structure of basic strings and a generative treatment of transitive and intransitive verbs*. Paper read before the 1966 Tokyo International Seminar in Linguistic Theory, Tokyo.

Havers, Wilhelm (1911). *Untersuchungen zur Kasussyntax der indogermanischen Sprachen*. Strasbourg, K. J. Trübner.

Heger, Klaus (1966). "Valenz, Diathese und Kasus." *Zeitschrift für romanische Philologie* **82**: 138–170.

Helbig, Gerhard (1971). *Beiträge zur Valenztheorie.* The Hague, Mouton.

Helbig, Gerhard and Wolfgang Schenkel (1969). *Wörterbuch zur Valenz and Distribution Deutsche Verben.* Leipzig, VEB Verlag Enzyklopädie.

Helbig, Gerhard and Wolfgang Schenkel (1973). *Wörterbuch zur Valenz und Distribution Deutscher Verben.* Leipzig, VEB Verlag Enzyklopadie.

Hjelmslev, Louis (1935, 1937). "La catégorie des cas: première partie and deuxieme partie." *Acta Jutlandica* **VII, IX**(1, 2).

Huddleston, Rodney (1970). "Some remarks on case grammar." *Journal of Linguistics* **1**: 501–510.

Ikegami, Yoshihiko (1970). *The Semological Structure of the English Verbs of Motion: A Stratificational Approach.* Tokyo, Sanseido.

Ivic, Milka (1962). "The grammatical category of non-omissible determiners." *Lingua* **11**: 199–204.

Ivic, Milka (1964). Non-omissible determiners in Slavic languages. *Proceedings of the Ninth International Congress of Linguists.* Horace G. Lunt. The Hague, Mouton: 476–479.

Jackendoff, Ray S. (1972). *Semantic Interpretation in Generative Grammar.* Cambridge, M.I.T. Press.

Jakobson, Roman (1936/1966). "Beitrag zur allgemeinen Kasuslehre: Gesamtbedeutungen der russuschen Kasus." *Travaux du Cercle Linguistique de Prague* **6**: 240–288.

Jakobson, Roman (1958a). Typological studies and their contribution to historical comparative linguistics. *Proceedings of the VIIIth International Congress of Linguistics.* Eva Sivertsen. Oslo, Oslo University Press: 17–25.

Jakobson, Roman (1958b). Morfologiceskie nabljudenija nad slavjanskim skloneniem. *In his Selected Writings, Vol. 11 (1971).* The Hague, Mouton: 154–183.

Jespersen, J. Otto (1924). *The philosophy of grammar.* New York, H. Holt and company.

Jespersen, J. Otto (1927). *A Modern English Grammar, vol. III.* London, Allen & Unwin.

Johnson, David (1974a). "On the role of grammatical relations in linguistic theory." *Papers from the 10th Regional Meeting of the Chicago Linguistic Society*: 269–283.

Johnson, David (1974b). "Toward a theory of relationally based grammar." *Ph.D. dissertation, University of Chicago.*

Katz, Jerrold J. (1964). "Semantic Theory and the Meaning of 'Good'." *Journal of Philosophy* **61**.

Katz, Jerrold J. (1972). *Semantic Theory.* New York, Harper and Row.

Katz, Jerrold J. and Jerry A. Fodor (1963). "The structure of a semantic theory." *Language* **39**: 170–210.

Kay, Paul and Linda Coleman (1978). "How to lie (with statistics)." *Ms.*

Keenan, Edward (1975). Some universals of passive in relational grammar. *Papers from the 11th Regional Meeting of the Chicago Linguistic Society.* L. James San Robin E. Grossman, Timothy J. Vance. Chicago, Chicago Linguistic Society: 340–52.

Keenan, Edward and Bernard Comrie (1972). "Noun phrase accessibility and universal grammar." *Ms.*

Kuipers, Aert H. (1962). "The Circassian nominal paradigm: A contribution to case theory." *Lingua* **11**: 231–248.

Kuno, Susumu (1973). *The Structure of the Japanese Language.* Cambridge, M.I.T. Press.

Kuryłowicz, Jerzy (1949). Le problème du classement des cas. *In his Esquisses Linguistigues (1960).* Wroclaw–Krakow, Ossolineum.

Kuryłowicz, Jerzy (1960). Le problème du classement des cas. *Esquisses linguistiques.* Wroclaw-Krakov, Zaklad Naroldowy im. Ossolínskich: 131–150.

Kuryłowicz, Jerzy (1964). *The inflectional categories of Indo-European.* Heidelberg, C. Winter.

Labov, William (1973). The boundaries of words and their meanings. *New Ways of Analyzing Variation in English.* Charles-James Bailey and Roger Shuy. Washington D. C., Georgetown University Press: 340–373.

Lakoff, George (1965). On the Nature of Syntactic Irregularity. Cambridge, Mathematical Linguistics and Automatic Translation, Harvard Computation Laboratory.

Lakoff, George (1966). Stative adjectives and verbs in English. Cambridge, The Computation Laboratory of Harvard University Mathematical Linguistics and Automatic Translation: I-1 to I-16.

Lakoff, George (1967). Instrumental adverbs and the concept of deep structure. *Duplicated.* Cambridge.

Lakoff, George and Stanley Peters (1966). Phrasal conjunction and symmetric predicates. Cambridge, The Computation Laboratory of Harvard University Mathematical Linguistics and Automatic Translation: VI-1 to VI-49.

Lane, George S. (1951). "Review of Y.M. Biese, Some notes on the origin of the Indo-European nominative singular." *Language* **27**: 372–374.

Langendoen, D. Terence (1966). Some problems concerning the English expletive 'it', The Ohio State University Research Foundation Project on Linguistic Analysis: 104–134.

Lee, P. Gregory (1967). "Some properties of 'be' sentences." *Ms.*

Leech, Geoffrey (1974). *Semantics.* London, Penguin.

Lehmann, Winfred P. (1958). "On earlier stages of the Indo-European nominal inflection." *Language* **34**: 179–202.

Leisi, Ernst (1952). *Der Wortinhalt.* Heidelberg, Quelle and Meyer.

Lévy-Bruhl, Lucien (1916). "L'espression de la possession dans les langues mélanésiennes." *Bulletin de la Société de linguistique de Paris* **19**: 96–104.

Lotz, János (1939). *Das Ungarische Sprachsystem.* Stockholm, Ungarisches Institut.

Lyons, John (1966). "Towards a 'notional' theory of the 'parts of speech'." *Journal of Linguistics* **2**: 209–236.

Manessy, Gabriel (1964). La relation génitive dans quelques langues mandé. *Proceedings of the Ninth International Congress of Linguists.* Horace G. Lunt. The Hague, Mouton: 467–475.

Martin, W. (1972). *OWL, Automatic Programming Group.* Cambridge, M.I.T. Press.

Martinet, André (1962a). *A functional view of language.* Oxford, Clarendon Press.

Martinet, André (1962b). "Le sujet comme fonction linguistique et l'analyse syntaxique du Basque." *Bulletin de la Société de linguistique de Paris* **57**: 72–83.

McCawley, James D. (1971). "Prelexical syntax." *Georgetown University Monograph Series on Languages and Linguistics* **24**.

McKaughan, Howard (1962). "Overt relation markers in Maranao." *Language* **38**: 47–51.

Meinhof, Carl (1938). Der Ausdruck der Kasusbeziehungen in afrikanischen Sprachen. *Scritti in onore di Alfredo Trombetti.* Ulrico Hoepli. Milan: 71–85.

Mellema, Paul (1974). "A brief against case grammar." *Foundations of Language* **11**: 39–76.

Miller, George A. (1972). English verbs of motion: a case study in semantics and lexical memory. *Coding Processes in Human Memory.* A. W. Melton and E. Martin. Washington D. C., V. H. Winston.

Minsky, Marvin (1975). A framework for representing knowledge. *The Psychology of Computer Vision.* Patrick H. Winston. New York, McGraw-Hill.

Müller, C. F. W. (1908). *Syntax des Nominativs und Akkusativs im Lateinischen.* Leipzig, B. G. Teubner.

Newmark, Leonard (1962). "An Albanian case system." *Lingua* **11**: 312–321.

Nida, Eugene (1975). *Componential Analysis of Meaning.* The Hague, Mouton.

Nilsen, Don L. F. (1972). *Toward a Semantic Specification of Deep Case.* The Hague, Mouton.

Oertel, Hanns (1936). *The syntax of cases in the narrative and descriptive prose of the Brahmanas.* Heidelberg, C. Winter.

Paige, Jeffrey M. and Herbert A. Simon (1966). Cognitive processes in solving algebra word problems. *Problem Solving: Research, Method and Theory.* Benjamin Kleinmutz. New York, John Wiley & Sons.

Panevová, Jarmila (1974). "On verbal frames in functional generative description, Part 1." *Prague Bulletin in Mathematical Linguistics* **No. 22**.

Perlmutter, David and Paul Postal (1974). "Linguistic Institute lectures on relational grammar." *Ms.*

Pike, Kenneth L. (1966). Tagmemic and matrix linguistics applied to selected African languages. Washington D.C., U.S. Department of Health, Education, and Welfare, Office of Education, Bureau of Research: esp. Chapter 1, Clauses.

Postal, Paul M. (1964). Mohawk prefix generation. *Proceedings of the Ninth International Congress of Linguists.* Horace G. Lunt. The Hague, Mouton: 346–355.

Postal, Paul M. (1966). On so-called 'pronouns' in English. *Report of the Seventeenth Annual Round Table Meeting on Linguistics and Language Studies (identical with Georgetown University Monograph Series on Languages and Linguistics 19).* Francis P. Dinneen. Washington D.C., Georgetown University Press: 177–206.

Pullum, Geoffrey K. (1977). Word order universals and grammatical relations. *Syntax and Semantics 8: Grammatical Relations.* Peter Cole and Jerrold M. Sadock. New York, Academic Press: 249–277.

Raible, Wolfgang (1975). Partitiv, Akkusativ und Nominativ als Objekt. *Proceedings of Bielefeld Conference on Case Labels, April 1975.* János Petöfi. Bielefeld, University of Bielefeld.

Redden, James E. (1966). "Walapai II: Morphology." *International Journal of American Linguistics* **32**: 141–163.

Robins, Robert H. (1961). "Syntactic analysis." *Archivum linguisticum* **13**: 78–89.

Rohdenburg, Günter (1970). Zum persönlichen Subjekt im Englischen. *PAKS Arbeitsbericht No. 6, Lehrstuhl Anglistik.* Stuttgart, University of Stuttgart: 133–164.

Rosch, Eleanor (1973). "Natural categories." *Cognitive Psychology* **4**: 326–350.

Rosén, Haiim (1959). "Die Ausdrucksform für 'veräusserlichen' und 'unveräusserlichen' Besitz in Frühgriechischen." *Lingua* **8**: 264–293.

Rosenbaum, Peter (1968). *The Grammar of English Predicate Complement Constructions.* Cambridge, M.I.T. Press.

Ross, John R. (1970). On declarative sentences. *Readings in English Transformational Grammar.* Roderick A. Jacobs and Peter S. Rosenbaum. Waltham, Ginn & Blaisdell.

Saint-Jacques, Bernard (1966). *Analyse structurale de la syntaxe du japonais moderne.* Paris, C. Klincksieck.

Salzmann, Zdenek (1965). "Arapaho VI: Noun." *International Journal of American Linguistics* **31**: 136–151.

Sankoff, Gillian and Johanne Thibault (1978). "Sacrilege in Quebec French." *Ms.*

Sapir, Edward (1917). "Review of C. C. Uhlenbeck, 'Het Identificeerend karakter der possessieve flexie in talen van Noord-Amerika'." *International Journal of American Linguistics* **1**: 86–90.

Schank, Roger (1975). The structure of episodes in memory. *Representation and Understanding: Studies in Cognitive Science.* Daniel G. Bobrow and Allen Collins. New York, Academic Press.

Sommerfelt, Alf (1937). Sur la notion du sujet en géorgian. *Mélanges de linguistique et de philologie offerts à Jacques van Ginneken.* Paris, C. Klincksieck: 183–185.

Staal, J. Frits and Yehoshua Bar-Hillel (1967 and 1968). "Foundations of Language."

Strawson, Peter F. (1950). "On referring." *Mind* **50**: 320–344.

Talmy, Leonard (1972). "Semantic Structures in English and Atsugewi." *Unpublished doctoral dissertation, University of California.*

Tesnière, Lucien (1959). *Éléments de syntaxe structurale.* Paris, Klincksieck.

Trubetzkoy, Nikolai S. (1939). Le rapport entre le déterminé, le déterminant et le défini. *Mélanges de linguistique offerts à Charles Bally.* s.a. Georg et cie. Geneva: 75–82.

Uhlenbeck, C. C. (1901). "Agens und Patiens im Kasussystem der indogermanischen Sprachen." *Indogermanische Forschungen* **12**: 170–171.

Ullmann, Stephen (1952). *Précis de Sémantique Française.* Berne, A. Francke.

Ullmann, Stephen (1958). Discussion in the 'Structure of Meaning' session. *Proceedings of the Eighth International Congress of Linguistics.* Eva Sivertsen. Oslo, Oslo University Press.

Vaillant, André (1936). "L'ergatif indo-européen." *Bulletin de la Société de linguistique de Paris* **27**: 93–108.

van Ginneken, Jacques (1939). Avoir et être du point de vue de la linguistique générale. *Mélanges de linguistique offerts à Charles Bally.* Geneva, Georg et cie, s.a.: 83–92.

Velten, H. V. (1962). "On the functions of French de and à." *Lingua* **11**: 449–452.

Weniger, Dorothea (1974). "The Semantic Structure of English Motion Verbs." *University of Zurich Ph.D. Dissertation.*

Whorf, Benjamin Lee (1965). A linguistic consideration of thinking in primitive communities (c. 1936). *Language, thought and reality: Selected writings of Benjamin Lee Whorf.* John B. Carroll. Cambridge, M.I.T. Press: 65–86.

Wierzbicka, Anna (1972). Emotion. *Semantic Primitives.* Frankfurt, Athenäum: 57–70.

Wierzbicka, Anna (1980). *The Case for Surface Case: Linguistica Extranea Studia 9.* Ann Arbor, Karoma.

Zsilka, János (1967). *The System of Hungarian Sentence Patterns.* Bloomington, Indiana University Press.

Author Index

U

V

W

Z

Subject Index

Word Index